Humboldt *and* Jefferson

Humboldt

A TRANSATLANTIC FRIENDSHIP

and Jefferson

OF THE ENLIGHTENMENT

Sandra Rebok

University of Virginia Press *Charlottesville and London*

University of Virginia Press

Printed in the United States of America on acid-free paper

First published 2014

9 8 7 6 5 4 3 2 1

LIBRARY OF CONGRESS CATALOGING-IN-PUBLICATION DATA
Rebok, Sandra.
Humboldt and Jefferson : a transatlantic friendship of the enlightenment / Sandra Rebok.
pages cm
Includes bibliographical references and index.
ISBN 978-0-8139-3569-0 (cloth : alk. paper) — ISBN 978-0-8139-3570-6 (e-book)
1. Humboldt, Alexander von, 1769–1859. 2. Jefferson, Thomas, 1743–1826. 3. Scientific expeditions. I. Title.
Q143.H9R3625 2014
509.2—dc23

2013041208

Contents

Acknowledgments

My extensive scholarly research on Alexander von Humboldt, which I began seventeen years ago, has opened my eyes to numerous fascinating historical topics, beautiful geographic regions, and interesting personalities. Without a doubt, Thomas Jefferson was one of the most intriguing persons Humboldt met during his American expedition (1799–1804) and corresponded with for more than twenty years. This is partly due to Jefferson's personality, his manifold interests and achievements, but also to the fact that the transatlantic exchange of ideas and knowledge they sustained until Jefferson's death in 1826 touched upon many of the eminent issues of their time. I am therefore grateful for having had the opportunity to become immersed in the long-lasting relationship of these two captivating minds through my research at the Alexander von Humboldt Research Center in Berlin as well as the Robert H. Smith International Center for Jefferson Studies and the University of Virginia in Charlottesville. The opportunity to study Jefferson's and Humboldt's writings, and the vast secondary literature connected to the issues they raised in their communication, while being at historical places that retain something of their spirit—in Berlin, Paris, Philadelphia, Washington, or overlooking the beautiful landscape of Monticello, Jefferson's estate in Charlottesville—has been truly inspiring.

My deep appreciation, therefore, goes to the Robert H. Smith International Center for Jefferson Studies and the Thomas Jefferson Foundation, as well as to the Deutsche Forschungsgemeinschaft, for granting me several research fellowships that were instrumental in the preparation of this book; and also to the Spanish Ministerio de Economía y Competitividad for the economic support received from the research project "Naturalists and Travelers in the Hispanic

World: Institutional, Scientific and Teaching Aspects" (HAR2010-21333-C03-02). In addition, a number of colleagues have been important to the development of this work. First, I would like to mention the support that James R. Sofka, Miguel Ángel Puig-Samper, and Andrew O'Shaughnessy have provided at different levels from the earliest stages of this project. Furthermore, I would like to thank Boyd Zenner from the University of Virginia Press for her interest in the relationship between Humboldt and Jefferson, as well as Pilar Tigeras at the Spanish National Research Council for facilitating the necessary research trips. I am also grateful to Aurelio Hinarejos, who gave me his much-appreciated support in all phases of the preparation of this work, from the first research undertaken to the final revisions of the manuscript.

Moreover, I want to thank my dear colleagues at the International Center for Jefferson Studies, in particular Anna Berkes, Endrina Tay, Gaye Wilson, Lisa Francavilla, and Jack Robertson, who have always shared their knowledge with me during my research stays there as well as through e-mail correspondence. My conversations with other Jefferson scholars working at this research center have been very inspiring as well; I am thus grateful for all the Fellows with whom I was able to discuss my research topics, among them Michael Kranish, James Thompson, Katherine Woltz, Doug Bradburn, Charlene Boyer Lewis, and Carrie Douglas. My acknowledgments would not be complete without including my other Virginian friends who have made my visits to Charlottesville a very pleasant experience.

In a more advanced stage of this work, the first readers of my book manuscript contributed with numerous valuable suggestions. This refers also to the recommendations of those who evaluated the articles on the relationship between Humboldt and Jefferson that I had previously published in the *Southern Quarterly*, the *Virginia Magazine of History and Biography*, *French Colonial History*, and in a chapter of the book *Bernhard Varenius (1622–1650)*. Their comments, based on their broad scholarly expertise, helped to improve my manuscript, and I extend my thanks also to them. Finally, I would also like to express my appreciation of the work that Mark Mones and Susan Murray have done in editing and preparing the text for publication.

Last but not least, the research and preparation of a book manuscript depend on ideas received through other publications. I therefore acknowledge the importance of this source of inspiration for my work, hoping that this book, too, may contribute to the work of other scholars.

Humboldt *and* Jefferson

Introduction

The transfer of ideas, impressions, and knowledge among those traveling between the Old and New Worlds was particularly vital at the end of the eighteenth and the beginning of the nineteenth century, a period characterized by the questioning of the traditional understanding of the structure of the world and by the search for a new social order. These lines of inquiry flowed directly from the Enlightenment. During the eighteenth century, many intellectuals advocated turning away from a reliance on tradition and religious belief and proposed to reform society through the use of rational principles, and to advance knowledge through science and intellectual exchange. In 1784, Immanuel Kant published "Answering the Question: What Is Enlightenment?" (Beantwortung der Frage: Was ist Aufklärung?) in the *Berlinische Monatsschrift*.[1] In this well-known essay, Kant characterizes the Enlightenment as "man's emergence from his self-incurred immaturity," an immaturity self-inflicted because it arises not from a lack of understanding, but rather from an absence of the courage necessary to foreground reason and intellect. The Enlightenment was a crucial time not only for new political concepts but also for scientific thinking. Important Enlightenment figures lived on both sides of the Atlantic, and their correspondence formed a conduit through which ideas and information passed between the continents.

The Prussian traveler and scientist Alexander von Humboldt (1769–1859) and the American statesman, architect, and naturalist Thomas Jefferson (1743–1826) maintained a lively and productive transatlantic dialogue throughout their lives, and their writings about Europe and America have had a particularly long-lasting historical impact. Both of these cosmopolitan thinkers saw clearly the deficiencies of contemporary European society, and both believed that the United States held great promise as the model for a better society. They recognized the immense importance of creating an international network through which correspondence might flow multidirectionally, addressing the most significant questions and publications of the time. Such scholarly communities, Jefferson wrote, "are always in peace, however their nations may be at war. Like a republic of letters they form a great fraternity spreading over the whole earth, and their correspondence is never interrupted by any civilized nation."[2]

Alexander von Humboldt's visit to the United States and his initial encounter with Thomas Jefferson took place in spring of 1804, when he had finished his scientific expedition through the Spanish colonial territories in America. At the time of his travels, these were divided into the viceroyalties of New Spain, New Granada, Peru, and Cuba. Given the cultural significance of both men, their meeting and subsequent friendship, expressed through correspondence over the following twenty years, has been a strong point of interest for many scholars.[3]

Humboldt and Jefferson's personal and ideological affinities still resonate today, particularly in the fields of intellectual and Atlantic history, and the history of science. Their exchanges touched the pivotal events of their time, including the independence movement in Latin America, the applicability of the democratic model to that region, the relationship between America and Europe, and the development of diverse technological projects. Both were supporters of the Enlightenment principles that led to the French Revolution, and both initially held hopes for the impact of the movement in France and elsewhere, even while they deplored its bloody repercussions. Nevertheless, the two men occupied different sociopolitical worlds, and these influenced both their rhetoric and their actions.

As a starting point, it is useful to compare the ways in which Humboldt and Jefferson were influenced by their transatlantic experiences. Jefferson's conception of the future of the United States was strongly marked by his tenure in France as minister plenipotentiary (1784–89).[4] Similarly, the five years Humboldt spent among colonial societies and his visit to the first free nation in America gave him a new perspective on politics. While their experience abroad bred ambivalence in both men, it also molded their convictions and in turn shaped their personal and political ideas.

Both staunch advocates of the ideals of the Enlightenment behind American independence, Jefferson and Humboldt saw in the young democracy the political system of the future. Their views differed in several important ways, however, as did their general understanding of the postulates of the Age of Reason and their opinions on how best to put these precepts into practice. For example, their responses to the Haitian Revolution (1791–1804), the slave revolt in the French colony of Saint-Domingue, highlighted their views on race as well as on social revolution as a means of restructuring society. Nevertheless, both were scholars with a global perspective who had a marked interest in the advancement of science and the exploration of the New Continent, and who argued vigorously against the much-debated theory that assumed American inferiority. Both also assigned great weight to the understanding of and interaction with the natural world, and their contributions to the broad field of natural history demonstrate an early involvement with what would come to be seen as important environmental concerns.

1 Biographical Backgrounds

Alexander von Humboldt

Alexander von Humboldt was born on September 14, 1769, in a small palace in the town of Tegel, near Berlin. He spent his childhood with his older brother, Wilhelm,[1] with whom he maintained a close relationship throughout his life.[2] The two boys were raised in an aristocratic family. Their father, Alexander Georg von Humboldt, was chamberlain to the Prussian king and an important figure at court. Their mother, Marie Elisabeth von Humboldt (née Colomb), was a wealthy woman who had decisive impact on the young Alexander. The Humboldts engaged as their sons' tutor a well-known writer and linguist, Joachim Heinrich Campe, who strongly influenced the intellectual development of the Humboldt brothers, as did another teacher, Gottlob Johann Christian Kunth (1757–1829), who encouraged them to study languages. Campe and Kunth contributed substantially to the brothers' success in the cultural circles of the time. The salons of the Jewish community in enlightened Berlin, particularly that of Marcus Herz and his wife, Henriette, were privileged cultural and social enclaves that also played an important role in Alexander's education.

Humboldt was a child of the German Enlightenment, which had its inception in the 1780s in reaction to Kant's *Kritik der reinen Vernunft* (*Critique of Pure Reason*) and lasted until the death of Georg Wilhelm Friedrich Hegel in 1831.[3] Alexander's formal education brought him in direct contact with Enlightenment ideals, first in his native city and then during his studies at the universities of Frankfurt and Göttingen. He continued his studies for one year at the Academy of Commerce in Hamburg and after that at the Academy of Mining in Freiberg, Saxony, where he was taught by the eminent geologist Abraham Gottlob Werner. In 1792, Humboldt was appointed to the position of assessor in the mining department. Shortly afterward he was promoted to the post of superior mining officer in the Franconian principalities.

These appointments marked the beginning of Humboldt's serious study of mineralogy and natural history. Previously he had traveled through the Netherlands, England, and France with the famous naturalist Georg Forster, who, with his father, Reinhold, had accompanied Captain James Cook on his second expedition around the world. Humboldt dedicated several years to mining, but during this period he also published an encyclopedia of Freiberg flora,[4] as well as several monographs on physics and chemistry, some of which were published in French and British journals.

Humboldt came of age in an era of great explorations, such as the voyages undertaken by Louis Antoine de Bougainville, Jean-François de La Pérouse, James Bruce, Carsten Niebuhr, and Alejandro Malaspina and José de Bustamante; or those carried out by James Cook. The descriptions of their adventures had fascinated Humboldt from his early youth and formed his image of the tropical realm idealized by the philosopher Jean-Jacques Rousseau. He devoured the works of Haller, Macpherson, and Goethe that imagined the return of human beings to their original state, far from civilization. Humboldt learned much about what were to him exotic worlds through the works of the French writer and botanist Jacques-Henri Bernardin de Saint-Pierre—whose novel *Paul et Virginie* he read repeatedly—as well as those of his preceptor Campe, author of *Robinson, der Jüngere* (1779) and *Die Entdeckung Amerikas* (1781–82), which made a particularly strong impression on him. Although the books did not provide him with much concrete information on obscure and distant territories, they awoke in him a fierce desire to experience these faraway and alluring lands himself. From his early youth he longed to undertake a real scientific expedition.

Another significant influence upon the young Humboldt was the Prussian pharmacist and plant taxonomist Carl Ludwig Willdenow, who became the most important botanist in Berlin. In 1798, he was made professor for natural history at the Collegium medico-chirurgicum, and three years later he was appointed botanist at the Academy of Sciences. From 1801 until his death he directed the

Botanical Garden, and after the Prussian king Friedrich Wilhelm III created the Friedrich-Wilhelms-Universität in Berlin in 1809, Willdenow was named professor of botany. By this time, his first work, *Florae Berolinensis prodromus,* had been very well received in the scholarly community, which had led to a correspondence between Humboldt and Willdenow. Humboldt visited Willdenow regularly in Berlin, where the older man instructed him in botany, and particularly on cryptogams. Humboldt based his early botanical studies on Willdenow's *Florae Berolinensis prodromus,* which also served as the inspiration for Humboldt's own first botanical work, *Florae fribergensis specimen,* dedicated to his mentor.

Humboldt's precocity and scholarly achievements attracted both national and international notice. Thus, when the French government decided in 1798 to undertake a circumnavigation of the globe directed by Nicolas Thomas Baudin,[5] Humboldt was invited to join the group of naturalists on board. While waiting for that expedition to start, the young Prussian continued his scientific work in Paris, where he became acquainted with the French botanist Aimé Bonpland,[6] who had also been invited on the Baudin voyage. When the expedition was canceled for financial reasons, Humboldt—having fortuitously received an inheritance from his parents—decided to mount his own voyage of exploration and was able to convince Bonpland to join him. After several failed attempts to initiate their scientific journey elsewhere, Humboldt and Bonpland finally went to Spain, hoping to undertake an expedition to the Spanish colonies overseas. They traveled through the Iberian Peninsula from January to May 1799, a period that proved critical to the scientific preparation as well as the diplomatic approval of their project. Humboldt needed the permission of King Carlos IV to carry out his scientific voyage through Spanish dominions, something rather difficult to obtain, since the court in Madrid—like other European powers—preferred to keep foreign travelers out of their colonial territories. Nevertheless, Humboldt's intellect and accomplishments impressed the king, who granted the travelers an unprecedented and unrestricted permission to undertake the planned voyage.

Madrid proved ideal for the preparation of such a scientific expedition, since participants in previous Spanish expeditions to America were readily available and Humboldt could study their New World natural-history collections. He established close contacts with the naturalists in scientific institutions such as the Real Jardín Botánico and the Real Gabinete de Historia Natural, particularly with the leading Spanish botanist, Antonio José Cavanilles, and the writer José Clavijo y Fajardo, as well as with the naval officer and geographer José Espinosa y Tello and the historian Juan Bautista Muñoz at the Depósito Hidrógrafico and the Real Academia de la Historia, respectively. He also met with the German specialists in the scientific community of Madrid, among them Christian Herrgen, Johann Wilhelm and Heinrich Thalacker, and the Heuland brothers. In spite of his relative youth, Humboldt had already embarked upon a brilliant scientific career and thus was an interesting contact for the scholars in Madrid. He had been working for the Prussian state for five years as an expert on mines, and was the author of several scientific publications.[7] Humboldt and Bonpland made good use of the time they spent in Spain, taking measurements during their travels through the peninsula and testing the new scientific instruments they had brought from Paris.

At the beginning of June 1799, Humboldt and Bonpland left from La Coruña in northwest Spain on the corvette *Pizarro* and headed for the Canary Islands, where they stayed for six days on Tenerife. They used this time to travel around the island and undertake extensive scientific studies, including climbing the highest mountain of Spain, the Teide, both in connection with their interest in vulcanism and to collect information for Humboldt's geography of plants.[8] From Tenerife, they set sail at last for the New World on the expedition that would cement Humboldt's global reputation.

They reached Cumaná, Venezuela, their first port in the Americas, on July 16, 1799. There Humboldt visited a mission at Caripe and explored the Guácharo cavern, where he encountered the oilbird (*Steatornis caripensis*) and became the first to describe this species. Back in Cumaná, he witnessed a remarkable meteor shower of the Leonids, and his observations later helped to explain the periodic character of this celestial event. After several other excursions to nearby places, Humboldt and Bonpland traveled to Caracas, and in February 1800, they initiated their first big expedition into the interior of the American continent to explore the course of the Orinoco River. During four months and accompanied by a group of Indians, they traveled through 1,725 miles of wild and largely inhabited country. They first descended the Apure River to the Orinoco River, then traveled on the Orinoco and later on the Atapabo River to the south, in order to reach the sources of the Negro River. From there they finally arrived at the Casiquiare River and were thus able to demonstrate the existence of a linkage between the water system of the Orinoco and Amazon Rivers through the Casiquiare. On May 20, 1800, they reached the bifurcation of the Orinoco and became the first to determine its exact position. In addition, they documented the life of several native tribes such as the Maipures and their already-extinct rivals, the Atures. On their way back, they followed the course of the Orinoco in the direction of Angostura (Ciudad Bolívar). Passing the Llanos in great heat, they continued to travel north until, on July 23, they approached Nueva Barcelona, a coastal town where they stayed until November 24, when they embarked for Havana, Cuba. Besides pursuing their extensive scientific interests in Cuba, Humboldt and Bonpland, after having spent months of traveling in the tropics under difficult circumstances, also enjoyed the social life in Havana. In March 1801, they sailed from Batabanó in southwestern Cuba to Cartagena de Indias, Colombia. Humboldt, having learned that the French captain Nicolas Baudin had finally been able to initiate his circumnavigation, hoped to meet with his expedition on the Peruvian coast. This decision also gave them the opportunity to explore the Andes. From Barancas Nuevas, their itinerary led them for forty days up the Magdalena River, passing Honda and Santa Fe de Bogotá, where they were well received by the Spanish botanist José Celestino Mutis, with whom they discussed their botanical discoveries. While there, Humboldt also prepared an expert's report for the Spanish viceroy about the silver mines and the gold production of Colombia. After a tedious journey, starting from Bogotá on September 19, crossing the Cordillera Real and with a short stay in Popayán, they arrived in Quito on January 6, 1802. The marquis of Selva Alegre, Juan Pío Montúfar y Larrea, accommodated them in his house, and his son Carlos Montúfar decided to accompany the group for the remaining part of their expedition.[9]

In Ecuador, Humboldt pursued his interest in volcanoes, ascending the Pichincha several times and attempting to reach the peak of the Chimborazo. A crevice in the rocks kept them from ascending higher than 19,286 feet, shortly below the peak, but they nevertheless established a world record that lasted for thirty years, a remarkable accomplishment considering that they lacked the necessary equipment for these heights and suffered from altitude sickness. They then mounted an expedition to the sources of the Amazon, exploring the upper course of the Marañón River, and then searched for the remaining parts of the Inca settlement near Cajamarca. They crossed the Andes again, arriving in Lima on October 23, 1802, where Humboldt observed the transit of Mercury on November 9 and determined the precise longitude of the city. Humboldt also studied and described the fertilizing properties of guano, creating considerable interest among Europeans in importing this product.

In the meantime, even before leaving Quito, Humboldt had discovered that Captain Baudin had modified his itinerary so that it would not be not possible to join his expedition. Humboldt then decided to embark first to Guayaquil, where during a short stay he was able to determine through measurements the ocean current now known as the Humboldt Current, and from there he sailed to Acapulco. With his arrival in New Spain after a tempestuous voyage on March 23, 1803, the final part of his expedition began. He lived for one year in this country, visiting different places on their way from Acapulco to Mexico City and from there to Veracruz on the Caribbean coast. Humboldt's basic interest was centered on the mining industry, and he visited the mines of Morán, Real del Monte, and Cerro del Oyamel. In Mexico City, he assisted at the exams in the Colegio de Minería, whose director and founder, Fausto de Elhuyar, he knew from his time in Freiberg. Humboldt also spent a large amount of time in Mexico City's colonial archives gathering statistical, political, and historic material as well as social and economical data concerning New Spain. On March 7, 1804, the travelers left Veracruz for Havana, where in April Humboldt presented a mineralogical report at the Sociedad Económica de Amigos del País. From Cuba they initially intended to return to Europe and thus conclude their expedition, but instead they took the Spanish ship *Concepción* to Philadelphia and added five weeks in the United States to their journey. As we will see, this unplanned visit would assume a special importance in Humboldt's life.

Humboldt, it should be remembered, received no financing for his journey, which was dedicated exclusively to scientific study. Though he conducted the expedition with the formal approval of Madrid—he was even charged with sending mineralogical and botanical specimens to the major scientific institutions in the Spanish capital—his project was not connected to the political interests of Spain or any other European power. Humboldt was therefore able to pursue his own scientific objective—to take measurements of every natural component of the New World, including plants, animals, minerals, and climate so that he might understand them in context—without external interference. To cite only one outcome of Humboldt's investigations, approximately sixty thousand plant species—many of them previously unknown to Europeans—were identified and subsequently described in a great number of publications.[10] Since the American territories Humboldt visited were still under Spanish rule, he also had the opportunity to witness the European colonial system on the eve of its demise. Although his interest initially focused on all aspects of natural history, his works of this period also contain some critical commentary on the structures of colonial societies, often along with suggestions for reform that were undoubtedly influenced by the Age of Reason and the French Revolution.[11]

Upon his return to Europe, Humboldt settled in Paris, where he lived for the next twenty-two years, working with a number of French scientists to publish the results of his expedition. He maintained regular contact with friends such as Berthelot, Gay-Lussac, Arago, and Chateaubriand. His time in the French capital came to an end in 1827. Unable to maintain financial independence, he was forced to return to Berlin, where Friedrich Wilhelm III impatiently demanded his presence at court. Until a few years before his death, Humboldt served the king—and afterward his successor, Friedrich Wilhelm IV—as an advisor, court chamberlain, and diplomat. He was also a tutor to the Crown prince and a member of the Privy Council.

In April 1829, after British authorities refused to grant him permission for a long-desired expedition to India, where he had hoped conduct comparative studies between Asia and America, Humboldt initiated a second major expedition, this time to Russia, where officials hoped to avail themselves of his expertise in mining. Accompanied during his Russian travels by the mineralogist and chemist Gustav Rose, the zoologist Christian Gottfried Ehrenberg, and his own servant Johann Seifert, Humboldt was received with all honors at the imperial court in Saint Petersburg, where they spent three weeks. The party's itinerary then took them to Moscow, Kasan, Perm, Jekaterinburg, and the Ural Mountains, where the Prussian naturalist was charged with finding diamonds for the tsar. Later they went to Tobolsk and from there to the Altai Mountains and the Chinese frontier, from where they returned to Omsk and Miask on the way to Astrakhan, on the coast of the Caspian Sea. Humboldt made observations regarding the salt extraction at the lake of Elton, and visited the German settlements on the Volga. In early November, the returning expedition reached Moscow and then, on the 13th, Saint Petersburg, after having traveled across vast steppes, where they measured the air temperature and humidity, assessed variations in levels of magnetism, and calculated the geodesic position of the places they had visited. Their geological and mineralogical studies concentrated on finding the diamonds desired by the tsar. Besides its discovery of diamonds in the gold-washings of the Urals, this eight-month exploration also led to interesting scientific discoveries, such as the correction of the estimated height of the Central Asian plateau. Although Humboldt recorded the results of his exploration in his work *Asie Centrale*, this expedition never achieved the fame of his journey through America.

Humboldt died in 1859 at the age of eighty-nine while at work on the fifth volume of his final publication, the *Cosmos*. His vitality and enthusiasm hardly diminished, his memory unimpaired, he left behind a lavish assortment of geological, zoological, botanical, and ethnographic specimens, a great number of maps of previously unknown regions, and numerous publications resulting from his work. His opinions and convictions on the natural history and structure of societies in the New World are well documented, as are his perceptions of the progress of scientific research in general. Today most of his publications have been translated into several languages and are used as references by scientists and travelers.[12]

Alexander von Humboldt was perhaps the greatest public intellectual of the nineteenth century. Not only did he leave the world an immense repository of information, but he also passed on his methods to the researchers who followed him. He promoted international science and seeded so many fields with productive new ideas that historians of science refer to the era as "Humboldtian." In recognition of his groundbreaking contributions to the exploration of the American continent and to scientific progress, more places around the world are named after him—villages, towns, counties, streets, schools, rivers, bays, mountains, glaciers, parks, forests—than any other historical figure.[13]

Humboldt himself minimized his achievements. In a letter to his publisher Johann Georg von Cotta, written in the final years of his life, he mentioned that his important and original works were only three: the geography of plants in connection with what he termed the "*Naturgemälde*"[14] of the tropical world, the theory of the isothermal lines, and his observations on geomagnetism that resulted in the establishment of magnetic stations around the world, indicated by Humboldt himself.[15] Like Thomas Jefferson, Humboldt did not suspect the enormous, centuries-long impact he would have on both sides of the Atlantic.

Thomas Jefferson

Thomas Jefferson was born the third of ten children on April 13, 1743, in Shadwell, Virginia. He was the son of Peter Jefferson, a successful planter and surveyor in Albemarle County, and Jane Randolph, a member of one of Virginia's most distinguished families.[16] In 1752, he began attending a local school run by a Scottish Presbyterian minister, and between 1758 and 1760, he was educated in history, science, and the classics by Reverend James Maury near Gordonsville, Virginia. At age sixteen, Jefferson entered the College of William and Mary and first met the law professor George Wythe, who became his mentor. For two years, he pursued studies in mathematics, metaphysics, and philosophy, enthusiastically devouring the writings of the British empiricists John Locke, Francis Bacon, and Isaac Newton. In 1762, Jefferson began to read law under Wythe and passed his bar examination three years later. Until 1773, he was active as a lawyer, with a client list that included members of the Virginia's elite families, among them the Randolphs, his mother's family. He served in local government as a magistrate and county lieutenant, and represented Albemarle County in the Virginia House of Burgesses. In 1774, following the British Parliament's passage of the Coercive Acts (also known as the Intolerable Acts, a series of laws related to the British colonies in North America), Jefferson wrote a set of resolutions that were expanded into an essay called *A Summary View of the Rights of British America,* his first published work, which put forth the radical notion that the colonists had a natural right to govern themselves.

Having inherited at twenty-six a considerable landed estate from his father, Jefferson soon after began building Monticello, his beautiful mountaintop house within sight of Shadwell. Part of Jefferson's ongoing effort to create a neoclassical environment for himself based on the architectural principles of Andrea Palladio and the classical orders, Monticello demanded large cash outlays and started Jefferson down his long road into a lifetime of debt. In 1772, at age twenty-nine, Jefferson married the twenty-three-year-old widow Martha Wayles Skelton, with whom he lived happily for ten years until her death. Their marriage produced six children, only two of whom survived to adulthood; only their oldest daughter, Martha, lived beyond the age of twenty-five. Jefferson never remarried, and he maintained Monticello as his home throughout his life, steadily expanding and modifying it until 1809, when he finally stopped making changes.

One topic that occupied Jefferson during his entire lifetime was the institution of slavery. Jefferson himself was one of the largest slaveholders in Virginia, having inherited slaves from both his father and father-in-law. At any given time, he owned around two hundred, who were housed at Monticello, adjacent Albemarle County plantations, and Poplar Forest, his estate in Bedford County, Virginia.

Jefferson was a gifted writer, and as a member of the Continental Congress he was chosen in 1776 to draft the Declaration of Independence. Adopted on the Fourth of July, the document has been regarded ever since as a charter of American and universal liberties, proclaiming that all men are equal in rights, regardless of birth, wealth, or status; and that the government is the servant, not the master, of the people. After Jefferson left Congress, he returned to Virginia, where he served in the legislature. In 1779, at the age of thirty-six, he was elected governor of Virginia, a position he held for two years. During this time, he moved the state capital from Williamsburg to Richmond and continued to advocate for educational reforms at the College of William and Mary, including the nation's first student-policed honor code.

In 1780, the newly appointed secretary to the French minister in Philadelphia, François Barbé de Marbois, sent the governors and other dignitaries of each American state a list of twenty-two queries designed to provide him and his government with pertinent information on the American colonies. Jefferson, who had extensive knowledge of western lands from Virginia to Illinois, was one of only two of the many recipients who provided Marbois with a response. Written in 1781 and enlarged during 1782 and 1783, Jefferson's detailed response to Marbois's "Queries" would later become Jefferson's *Notes on the State of Virginia.* The book, initially printed for private distribution in Paris in 1785 and published in London two years later, takes the reader from a survey of the landscape and natural features of Virginia through a discussion of the state's social characteristics and scientific achievements.[17] The discussion also encompasses Virginia's history and ethnography, and it concludes with an examination of the state's economic importance and a consideration of the legal and historic tradition that established Virginia's cultural boundaries. Besides providing much data about Virginia, *Notes on the State of Virginia* details Jefferson's ideas concerning religious freedom, the separation of church and state, representative government versus dictatorship, and much else. The book, upon which much of its author's fame as a scientist-philosopher was based, is regarded as the most important scientific and political work written by an American before 1785 and a striking proclamation of Enlightenment ideas.

Notes on the State of Virginia also categorically rejects the European supposition of American inferiority. This theory had been energetically championed by the leading French naturalist Georges Louis Leclerc, Comte de Buffon (1707–1788), though similar opinions had been expressed by David Hume, Georg Wilhelm Friedrich Hegel, Abbé Guillaume-Thomas Raynal in his *Histoire philosophique et po-*

litique (1770), Abbé Cornelis de Pauw in *Récherches philosophiques sur les Américains* (1768), and William Robertson in *History of America* (1777). The theory was based on the primary proposition that fewer species of animals existed in the New World than in the Old, and that whenever the same or similar species were to be found in both Europe and America, the European animals were larger.[18] Between 1749 and 1789, thirty-six volumes of Buffon's famous work *Histoire naturelle, générale et particulière* were published, and another eight volumes were added after his death by Bernard Germain de Lacépède, the last appearing in 1804. Buffon's particular argument was that, as a result of living in a cold and wet climate, all species found in America were feeble, and any new species imported to the New World would produce weak offspring and soon succumb to its new environment. Extrapolating from this novel idea, Buffon concluded that Native Americans, too, were intellectually limited and lazy. Raynal and Pauw believed that Buffon's theory did not go far enough: American "inferiority" extended to the Europeans who settled in America, and to their descendants. Buffon—the first Frenchman elected to membership in the American Philosophical Society (1768)—had never traveled to the New World; he was an armchair scientist who used the information gathered by explorers to support and promote his own convictions and conclusions.

The theory of American "degeneracy" could stand only as long as it was not confronted with a rigorous empirical examination undertaken by persons who had acquired the necessary data themselves: either residents of the New Continent or Europeans who had gathered the data during their own explorations of America. Not surprisingly, the controversy became very popular in those years: it provoked much debate in contemporary newspapers, journals, and books, and in the salons of Europe and the manor houses of America. Thomas Jefferson realized the possible long-term consequences of Buffon's assertions in terms of trade or immigration, and thus how important it was for Americans to disprove them.

In May 1784, Congress appointed Jefferson minister plenipotentiary to France, and he later replaced Benjamin Franklin as minister to France. During this period, Jefferson became a serious student of European culture and undertook long travels throughout the Continent.[19] Always in search of interesting items and ideas to bring or send home to America, he acquired books, seeds and plants, statues and architectural drawings, scientific instruments, and information. Architecture and art particularly inspired him; he also greatly enjoyed the salon culture of Paris. He dined with many of the city's most prominent aristocrats—but also with the protagonists of the 1789 French Revolution, including the Marquis de Lafayette and the Comte de Mirabeau.

In September 1789, while traveling to America on home leave, Jefferson was offered the post of U.S. secretary of state under his fellow Virginian, President George Washington. He assumed the role in 1790 and began a tenure marked by his opposition to the pro-British policies of Alexander Hamilton. Jefferson retired to Monticello for the first time in late 1793 but continued to oppose the policies of Hamilton and Washington. Only three years later, as the presidential candidate of the Democratic Republicans, he became vice president after losing to John Adams by three electoral votes. In 1800, he defeated Adams and was elected president in the first peaceful transfer of authority from one party to another in the history of the young nation. Jefferson's presidency (1801–9) is remembered for several major achievements, among which the most notable, perhaps, was the purchase from France of the Louisiana Territory—the land west of the Mississippi River stretching to the Rocky Mountains. In 1762, France had ceded this territory to Spain, but with the secret Treaty of San Ildefonso in 1800, Napoleon Bonaparte had regained it for France. He envisioned a great French empire in the New World and hoped to use the Mississippi Valley as a food and trade center to supply Saint-Domingue, which was to be the heart of this domain. However, given the difficulty France had in maintaining control of the island, where slaves under the leader of the Haitian Revolution, François-Dominique Toussaint Louverture (1743–1803), had gained power, and facing new war with Great Britain, the country did not have enough troops to defend its holdings. Without Saint-Domingue, the value of Louisiana diminished for the French, and Napoleon needed funds to support his military ventures in Europe. Jefferson took advantage of this opportunity and sent James Monroe to France to negotiate a purchase, advising him: "You cannot too much hasten it, as the moment in France is critical. St. Domingo delays their taking possession of Louisiana, and they are in the last distress for money for current purposes."[20] Monroe's mission was successful, and in April 1803, Robert Livingston, Barbé Marbois, and Monroe signed the Louisiana Purchase Treaty in Paris. The acquisition of a territory of approximately 827,000 square miles, for which the United States paid a total of $15 million, removed the French presence from the United States, doubled the size of the country, and laid the foundation for its westward expansion.[21] Nevertheless, some expressed doubts. At the time, the borders of the new land were not yet firmly established, nor were the characteristics of the new territory known. Information on the western part of the continent was generally limited to what could be learned from trappers, traders, and explorers.

A milestone in the exploration of the country was the Lewis and Clark expedition (1804–6), also called the Corps of Discovery, the first transcontinental expedition to the Pacific coast undertaken by the United States.[22] This journey of exploration had political as well as scientific aims, for Jefferson had entertained ideas of scouting the American frontier region even before the Louisiana Purchase was made. After that acquisition, he appointed as leaders of the new expedition two Virginia-born military veterans of Indian wars—Meriwether Lewis as leader and William Clark as his partner. They explored the Louisiana Territory and beyond, producing a wealth of scientific and geographic information that ultimately made a material contribution to the European-American settlement of the West. The expedition had several specific objectives, including finding a water route across the continent to facilitate commerce, following and mapping the rivers, collecting scientific data, studying the flora, fauna, and geography of the unexplored regions, becoming acquainted with their native inhabitants, and determining how the territories might be put to economic use.

In his second term, Jefferson encountered more difficulties on both the domestic and foreign fronts. He is mostly remembered for his efforts to maintain neutrality in the midst of the conflict between Britain and France. In 1809, he was succeeded as president by his friend James Madison. During the last seventeen years of his life, Jefferson remained at Monticello and dedicated himself to the life of a gentleman farmer, pursuing various scientific, technical, and botanical interests. He was a major book collector with an enormous library, much of which he sold to the Library of Congress in 1815 after the British set fire to the Capitol, destroying its book collection.

In his final years, Jefferson embarked on his last great project: the founding of the University of Virginia. He believed that an educated public formed the cornerstone of a free republican society, and his intention was to create an institution of higher learning free from religious influence, where students could specialize in disciplines not yet offered at other universities. The University of Virginia's design expressed its founder's aspirations for both state-sponsored education and an agrarian democracy in the new republic. Jefferson was a proponent of classically inspired architecture, which he believed reflected American ideals. His idea of creating specialized units of learning is embodied in the configuration of his campus plan, which he called the "Academical Village."

By July 1825, Jefferson's health had begun to deteriorate, and less than a year later he was confined to his bed. On the Fourth of July 1826, on the fiftieth anniversary of the adoption of the Declaration of Independence and just hours before the death of John Adams, Jefferson died at the age of eighty-three. The epitaph that Jefferson himself wrote described him as the author of the Declaration of Independence and the Virginia statute on religious freedom, and as the father of the University of Virginia. His political service as U.S. president and vice president, governor of Virginia, and secretary of state went unmentioned. The achievements for which he hoped to be remembered reflect the essence of Thomas Jefferson: the lifelong pursuit of political freedom, religious freedom, and freedom of the mind through education.

2 Humboldt's Visit to the United States

The inspiration for Humboldt's visit to the United States came from the American consul in Cuba, Vincent F. Gray. While Humboldt was in Havana, Gray learned that the Prussian was in possession of documents relating to a region of New Spain still little known to the American government. In April 1804, Gray sent two dispatches introducing and recommending Humboldt to Secretary of State James Madison. He speculated that the explorer might be a source of useful documents and firsthand knowledge, adding politely: "I take leave to recommend him to your particular friendship and protection, during his stay in the United States. . . . [W]hile he remains in the said States, insure to him that attention and consideration to which [by virtue of] his character, he is so just entitled."[1] In another letter to Madison, this one from May 8, Gray notes that Humboldt "is now on his route to the city of Washington, and will have it in his power to give you much useful information relative to the country adjoining."[2] The American diplomat was particularly intrigued by the materials and maps from the Spanish colonial archives that contained hitherto-unknown information on the disputed borders between the United States and New Spain—information that was especially valuable given the recent purchase of the Louisiana Territory. The Spanish minister in Washington, Carlos Martínez de Irujo y Tacón, known as the Marquis de Casa-Irujo, was politically antagonistic to Jefferson, who at that moment was angling to procure Spanish West Florida for the United States. Under the circumstances, useful information was difficult to come by.[3]

Humboldt, for his part, wanted to meet Thomas Jefferson, the famous American scientist and philosopher who also happened to be president of the United States. Naturally, the combination attracted Humboldt. In Europe, he had experienced political figures rather as creators of obstacles to his scientific work. Now here was a head of state dedicated to democratically advancing the interests of his people, while at the same time sharing Humboldt's devotion to science and other intellectual pursuits.

More pragmatically, Humboldt was extremely active in matters of diplomacy and also might have taken into consideration the career advantages of having such an illustrious intellectual patron. After spending years in the New World and observing the various colonial societies, many of them displaying characteristics he publicly criticized, Humboldt was no doubt intrigued by the opportunity to familiarize himself with an independent country in America. Accordingly, he arranged to visit the United States between May 20 and June 30, 1804, traveling the Eastern Seaboard and meeting several times with the president and members of his cabinet.

Before their first personal encounter, Humboldt introduced himself by letter to Jefferson, expressing admiration and respect for his intellect, work, and liberal ideology.[4] During their subsequent meetings, he also complied with Jefferson's requests for intelligence and the latest geographic and statistical information on New Spain, material of great value to the Jefferson administration. A lifelong bond between the two men would develop as a result of that first brief meeting—a friendship marked by a lively exchange of ideas conducted through correspondence and other, more formal writings.

During his stay in the United States, Humboldt remained exclusively among the political and scientific elite. Accompanied by Aimé Bonpland and Carlos Montúfar, he arrived in Philadelphia on May 24 with two trunks full of manuscripts, plants, and other natural-history collections.[5] Philadelphia had been the temporary capital of the United States from 1790 to 1800, and at the turn of the century it was still the nation's financial and cultural center, as well as its second-largest city after New York. Here Humboldt spent his first days, having found accommodations at an inn on Market Street, near the harbor.[6] A significant portion of Humboldt's stay in Philadelphia centered around the American Philosophical Society, founded by Benjamin Franklin in 1743 and reorganized in 1769. Inspired by the Royal Society of London and considered the country's first learned society, the American Philosophical Society has played an important role in American cultural and intellectual life from its inception to the present day. Up until around 1840—and in spite of being a private organization—it assumed many of the functions of a national science academy, library, museum, and patent office. Considering the Society's international reputation, it is not surprising that it was the social and intellectual center of Humboldt's visit to Philadelphia. Here he established important contacts and received significant support. He expressed enthusiasm for the scientific activities of the Society, participated in its meetings, and gave a major lecture in the Philosophical Hall of Philadelphia to discuss his New World expedition. At the time, Jefferson was serving as president of the Society,[7] which numbered among its membership many other well-known cultural figures of the day—among them the naturalists Caspar Wistar Jr. and Benjamin Smith Barton; the mathematics professor Robert Patterson; and the philanthropist John Vaughan—who presented Humboldt to the scientific and artistic community of Philadelphia, including the artist, naturalist, and collector Charles Willson Peale and the physician Benjamin Rush. Humboldt and his travel companions were also guests at the house of Valentin de Foronda, the Spanish general consul to the United States in Philadelphia, who had been a member of the Society since 1802.[8] Furthermore, on June 21 Humboldt was invited by Caspar Wistar as honorary guest to the famous weekly organized social and intellectual gathering for the scientific elite, the so called "Wistar parties." Accompanied by Peale, the English physician Anthony Fothergill, Reverend Nicholas Collin, and the chemist James Woodhouse, Humboldt left Philadelphia on May 29 and made his way by coach to Washington—as of 1800 the capital of the country—with stops along the way in Chester, Wilmington, Elkton, Charlestown, Havre de Grace, and Baltimore. Before leaving, Peale had prepared some black-and-white profiles of Humboldt to present as gifts for the people they were to meet in Washington. The artist had executed these works in the Museum of Natural History, Peale's personal collection of paintings as well as botanical, zoological and archaeological specimens. Also known as Peale's "American Museum," this important collection was opened to the public in 1786 and became the first museum in the Western Hemisphere dedicated to natural history. In the 1790s, Peale's plan was to convert the place into a national museum, and he hoped to secure Humboldt's endorsement of the project, as we will see later.

The little band of travelers spent the first two weeks of June in Washington, where they visited the Capitol, the Library of Congress, and the Senate Chamber, as well as nearby Alexandria and George Washington's former home, Mount Vernon. It was probably on June 5 that Humboldt met Jefferson for the first time.[9] The Virginian, "according to his usual politeness received these gentlemen in the most friendly manner," and in order to be near to his guests, advised them to stay at the City Tavern.[10] Over the course of his stay, Humboldt also made the acquaintance of the secretary of state and future president, James Madison; the painter Gilbert Stuart; and the architect, inventor, painter, and physician William Thornton, all of whom invited the party of travelers to social events at their homes. Thornton accompanied Humboldt to Georgetown, where they visited the French and British ambassadors. Jefferson, wrote Peale in his diaries, "had known before the high reputation of the Baron as a Philosopher and ingenious observer of nature, that had traveled a great deal."[11] In 1801, a few years before their meeting, the surveyor Joseph Elgar Jr. had written to Jefferson about Humboldt's scientific work, calling Humboldt an "authority sufficient respectable."[12] In a letter Humboldt sent to Jefferson upon his departure, he confided that his primary reason for coming to the United States was to meet the man who—over the years and through his writings, ideas, and actions—had inspired in him such feelings of admiration.

As Peale recalled in his diaries, on June 4 the party was invited to the President's House[13] for a "very elegant dinner." Appreciatively, he notes that "not a single toast was given or called for, or Politicks touched on, but subjects of Natural History, and improvements of the conveniences of Life. Manners of the different nations described, or other agreeable conversation animated the whole company."[14] In *The First Forty Years of Washington Society*, published in 1906, Margaret Bayard Smith, a close friend of Jefferson and James Madison's wife, Dolley Madison, and a leading figure in Washington society of the time, relates an anecdote that demonstrates the friendly relationship that had been established between Humboldt and Jefferson: Humboldt, arriving at the President's House one day, was shown unannounced into the drawing room, where he found

Jefferson seated on the floor, surrounded by several of his young grandchildren so noisily engaged in play that Humboldt's entrance was not immediately noticed. When Jefferson saw Humboldt, he stood up and shook hands with him, saying, "You have found me playing the fool Baron, but I am sure to you I need make no apology."[15] Humboldt seemed to feel comfortable in Washington, perceiving himself to be among people with enlightened views similar to his own. Shortly before leaving the United States, he wrote to Madison that the days he spent in Washington were "the most delightful of [his] life."[16]

It should be noted here that the personal encounter between Humboldt and Jefferson took place in Washington rather than at Monticello, as has been reported in some other sources. This error is probably based on a comment found in an 1810 letter to Jefferson from Humboldt, who writes that in his thoughts he often returns to Monticello, where he pictures Jefferson in the peaceful shade of a magnolia tree.[17]

Benjamin Silliman, an early Humboldt expert, met the aged scientist in Berlin and wrote of a supposed visit to Monticello in his book *A Visit to Europe in 1851*. Humboldt, at the time eighty-two years old, might have confused Monticello with Mount Vernon, or Silliman himself may have made this error. In any case, the visit could not have occurred, since from May 13 to July 26, 1804, Jefferson was in Washington, only leaving for Monticello later, after Humboldt had already departed. Traveling on horseback or by carriage from the capital to Jefferson's home in Virginia—a distance of 125 miles—would have taken too long for Humboldt to include it in his itinerary. Finally, no document exists in which Humboldt mentions a stay at Monticello.[18]

On June 13, Humboldt, Bonpland, and Montúfar left Washington for Lancaster, Pennsylvania, to visit the botanist Gotthilf Heinrich Ernst Mühlenberg and the surveyor Andrew Ellicott.[19] They arrived on June 16 and two days later returned to Philadelphia, where they remained for another twelve days, making preparations for their return to Europe. On June 30, they finally left New Castle aboard the *Favorite*, arriving at the port of Bordeaux, France, on August 3. During their last days in Philadelphia, Peale painted a portrait of Humboldt, which today is in the collection of the College of Physicians of Philadelphia, and Secretary of the Treasury

Albert Gallatin provided him with some statistical information on the United States respecting the population, export of domestically grown articles, and navigation, as well as some printed documents on revenue and expenses that Humboldt had requested for use in his publications.[20]

Shortly before they were to embark, Humboldt appealed to James Madison for a special passport designed to protect them and their belongings on their return voyage to Europe. Sailing from Philadelphia to France meant venturing the risk that the British, who were searching American vessels for French property, could take "Citoyen" Bonpland prisoner. Montúfar, Humboldt thus requested in his letter to Madison, should not be mentioned at all; Bonpland should be listed as Humboldt's secretary rather than as a French citizen; and all property should be declared as Humboldt's own.[21] Madison issued the document immediately, requiring all U.S. armed vessels to allow the Europeans to pass "without hindrance," to provide them in case of need "all necessary aid and succour in their Voyage," and to respect them as promoters of "useful science."[22]

It appears from Humboldt's correspondence that he would have preferred to remain longer in the United States. In a letter to Madison written in Philadelphia, he expresses interest in seeing the United States again in a few years, when the way from Missouri to the Pacific Ocean would be open. Humboldt knew of the government's interest in exploring the western regions of the country, and he elaborates on the details of such an expedition, in which he might go as far north as Mount Saint Elias in Alaska and the Russian possessions. "With health and courage," he concludes, "all that can be carried out."[23] To Albert Gallatin, he voices regret at having to leave such a beautiful country, where, he felt, the progress of the human spirit and civil liberty presented such a brilliant spectacle.[24] To William Thornton, he expresses the hope that they would meet again someday in the United States, since "the country that extended to the west of the mountains offered a wide field for scientific exploration."[25] And in a letter to John Vaughan, he confesses a deep desire to roam one day over the western territories, in which project he saw in Jefferson as just the right man to support him.[26] To the president himself, however, Humboldt said none of this, though he was well aware that Jefferson's eyes were set on America's "continental destiny."

Still, Humboldt returned to Europe after only six weeks in the United States. At the end of his five-year American expedition, he felt he needed to begin work on the publication of his extensive findings. When, almost thirty years later, Humboldt had not yet returned to the United States, Madison delicately reminded him of his promise to do so. "There may be little hope now that a fulfilment of your original intention, would be compatible with the many interesting demands on your time elsewhere," Madison wrote. "I can only assure you therefore, that, on a more favourable supposition, you would no where be welcomed by more general gratulations than among the Citizens of the United States."[27]

Humboldt's new friends and colleagues in Philadelphia and Washington also regretted his return to Europe. As Thornton wrote to Vaughan shortly after Humboldt's departure: "I have not been for so many Years so much gratified as by the visit of the Wise Men of the East whom you sent to us. But I am sorry the interesting Baron has pocketed all South America. I wish he could have rested his Limbs awhile & published his works here.—The treasures of knowledge he has amassed are worth more than the richest gold mine."[28] In a letter to Jefferson, Benjamin Smith Barton referred to "the explorer of South America" as "one of the most intelligent and active philosophers of our times."[29] Reverend John Bachman, a Lutheran minister and natural scientist, recalled in his memoirs his meeting with Humboldt in Philadelphia when he was only fourteen years old. He describes the Prussian as having been at the very center of the society, always willing to answer every question with gentleness, friendliness, and kindness.[30]

Peale, who spent quite a bit of time at Humboldt's side and therefore was able to present important firsthand impressions, commented in his diaries: "The Baron spoke English very well, in the German dialect. Here I shall take notice that he possessed a surprising fluency of Speech, & it was amusing to hear him speak English, French and the Spanish Languages, mixing them together in rapid speech."[31] William Armistead Burwell, Jefferson's private secretary at the time of Humboldt's visit, remarked that "his presence at Washington attracted from Philadelphia all the men of science & learning," adding that "Jefferson appeared delighted with Humboldt & said he was the most scientific man of his age he had ever seen."[32] In a letter to his wife, Hannah, Albert Gallatin referred to the "exquisite intellectual treat from Baron Humboldt" and highlighted the "mass of natural, philosophical, and political information" Humboldt had gathered from his American expedition, "which will render the geography, productions, and statistics of those countries better known than those of most European countries." He added, "We all consider him a very extraordinary man, and his travels, which he intends publishing on his return to Europe, will I think, rank above any other production of the kind."[33]

Humboldt also made a significant impression on the women he met during his travels in the United States. Dolley Madison was quite enthusiastic about the "great treat in the company of a charming Prussian Baron von Humboldt." She wrote that "all the ladies say they are in love with him, notwithstanding his want of personal charms" and called him "the most polite, modest, well-informed and interesting traveller we have ever met."[34] Another female view was recorded in a letter by Margaret Bayard Smith to her sister-in-law, who also had "the singular pleasure of enjoying a great deal" of the company of this "charming man" and hoped for his return: "An enlightened mind has already made him an American, and we are not without hopes, that after having scratched his curiosity with travel he will spend the remainder of his days in the United States. This will be a great acquisition for I am sure I speak without exaggeration when I pronounce him one of the most learned men of the age. All his knowledge too is subservient to practical purposes, and he has a heart as ardently glows with a love of his fellow men as could be expected from the best principles and the finest feelings."[35] Years later, Smith published a detailed and astute description of his character and conduct:

> Baron Humboldt, formed not his estimate of men and manners, by their habiliments and conventionalisms, and refined as were his tastes, and polished as were his manners, he was neither shocked or disgusted, as was the case with the British Minister (Mr. Foster) by the old fashioned form, ill-chosen colours, or simple material of the President's dress. Neither did he remark the deficiency of elegance in his person, or of polish in his manners, but indifferent to these external and extrinsic circumstances, he easily discerned, and most highly appreciated the intrinsic qualities of the Philosophic Statesman. . . . He was most truly a citizen of the world, and wherever he went he

> felt himself perfectly at home. Under all governments, in all climates, he recognized men as his brother. Kind, frank, cordial in his disposition, expansive and enlightened in his views, his sympathies were never chilled, his opinions never warped by prejudice. The varieties of condition, of character, of costumes he met with among the nations he visited, were never subjected to the test of his own feelings and perceptions, but tried by the universal standard of abstract principles of utility, justice, goodness. His visits at the President's House, were unshackled by mere ceremony and not limited to any particular hour.[36]

On his way back to Europe, Humboldt wrote a short narrative of his American expedition that constitutes his first and only complete description of it. The American Philosophical Society had asked him for an account of his American travels, and Humboldt agreed, observing that since so much incorrect information had been circulated about the expedition, he wanted to publish his own description.[37] From New Castle he sent the text to John Vaughan, who translated the twenty-page document into English and published it in under the title "Original Communication—Supplementary" in the *Literary Magazine and American Register*.[38] On July 20, approximately three weeks after his departure, the American Philosophical Society elected Humboldt a member.[39] This recognition of his contributions to the advancement of science was the first of numerous honorary memberships that would be bestowed on him by scholarly and scientific societies throughout his life, which include the American Antiquarian Society (1816), New York Historical Society (1820), New York Literary and Philosophical Society (1822), American Academy of Arts and Sciences (1822), Lyceum of Natural History of New York (1827), Geological Society of Pennsylvania (1834), Rhode Island Historical Society (1838), Academy of Natural Sciences of Philadelphia (1842), American Ethnological Society (1843) and the American Geographical and Statistical Society (1856).[40]

For several reasons, Humboldt's visit to the United States must be regarded as a special part of his American project. As noted, this visit did not form part of the expedition he had originally planned. He makes no mention of his desire to visit the United States in his correspondence or travel journals; rather, he announces in several letters that he intends to return directly to Europe via Mexico and Cuba.[41] Neither his letters nor the annotations in his diaries, which deal primarily with his stay in Spanish America, mention the time he spent in the United States, his activities there, the persons he met, or his general observations and impressions.[42] This offers convincing evidence that Humboldt had not envisioned this visit as part of his exploration voyage; it also suggests that the United States did not figure in his considerations regarding the scientific interests and objectives of his American expedition. Here we can see an interesting parallel between Humboldt's visit to this country at the end of his expedition and his stay in Spain prior to his departure for America. Neither his travel through the Iberian Peninsula nor his visit to the Canary Islands was previously planned, and he seemed to pursue interests there different from those on which he focused in Spanish America. Nevertheless, the time he spent in Spain was essential for the scientific and diplomatic preparation of his exploration of America and helped to establish important contacts and scientific networks, some of which he maintained for decades.

Humboldt's diary entries describe his travels from Havana to Philadelphia and finish with a comment on May 22, 1804.[43] While these sources show him to be preoccupied during this voyage with taking scientific measurements such as air and water temperatures in different locations, the diary also reveals another motivation for his unplanned visit to the United States. After two passengers were killed during an interlude of extremely stormy weather on May 9, Humboldt feared for his life as well as the scientific fruits of his five-year expedition. Although he seems to question the necessity of the trip, he mentions in his diary that it was undertaken in order to save his manuscripts and collections from the "perfidious Spanish policies."[44] This comment suggests that despite the generous travel permission he had received from the Spanish government, he was nevertheless wary of its representatives.

Humboldt's last comments in these documents are his first impressions at their arrival in North America, made on May 18, when they were able to see the coast between False Cape and James Cape. His diary entries indicate that the landscape appeared rather uniform and sad after the vegetation on the tropical coasts of Acapulco, Cumaná, and Guayaquil. However, this impression changed the next day when they sailed up the Delaware, a beautiful and majestic river flanked on either side with villages and busy with many boats. Sailing past New Castle, he noted its small hills and the surrounding dense vegetation. Again, with his point of comparison being the tropics of Spanish America, he was intrigued by the chimneys and lightning rods on even the smallest houses. His annotations end when they arrived at a place he calls Lost Ridge,[45] with "a beautiful hospital amidst beautiful alleys," where they were supposed to endure the required quarantine. Humboldt was impatient at the loss of time involved and contacted Zaccheus Collins, a philanthropist and member of the Society of Friends in Philadelphia, for help. It was the first letter Humboldt wrote in the United States, and the fact that he received immediate support shows that he was already well known at the time of his arrival in the United States.

In the description of his expedition published in the *Literary Magazine and American Register,* Humboldt simply notes, without further detail, that he had journeyed from Cuba to France "by the way of Philadelphia." Since Humboldt's diaries and published works contain no other description of his activities in and impressions of the United States, the account of his visit derives mostly from the comments of those he met, particularly those in the Peale diaries. Humboldt does, however, make scattered references throughout his works to his American acquaintances and the data he received from them afterward. He used the statistical material he accumulated to good effect in his comparative studies between the United States and parts of Spanish America. His *Political Essay on the Kingdom of New Spain,* for example, contrasts Mexican and U.S. production and exportation volume, with Humboldt eager to prove that productivity would increase faster in a free society than in a colony.

That Humboldt left his scientific instruments in Mexico—and so was unable to take measurements of comparable complexity and detail for the remainder of the expedition—suggests that his stay in the United States was motivated more by political and ideological considerations than by scientific ones. He was very interested in meeting certain people and being known to them, and he openly conceded this. He had not come, he proclaimed, "to see your great Rivers and Mountains, but to become acquainted with your great men."[46] After five years of traveling in the Spanish possessions of America, Humboldt must have been eager to return to Europe, so his expanding his voyage just he was ready to leave the New World underscores the importance he placed on his visit to the United

States. He had met the Founding Fathers and architects of the first independent nation on the American continent, and he had seen for himself the functioning of the first republican institutions in the New World. This was the realization of ideals he passionately embraced.

3 Transatlantic Experiences

Humboldt and Jefferson each visited the other side of the Atlantic only once, and each remained in his respective "other" world for about five years. The two men lived during an age of inquiry, and of new definitions of European and American identity, and their transatlantic experiences decisively impacted their ideas and convictions for the rest of their lives.

Both men viewed the world across the ocean with ambivalence. Jefferson's European experiences inspired him as he labored with his countrymen to create a new society in America, for as he acknowledged in a letter to Edward Rutledge, "the best schools for republicanism are London, Versailles, Madrid, Vienna, Berlin etc."[1] For Humboldt, the United States offered a new political and economic model that was consonant with many of the values of the Age of Reason. Before leaving North America, he wrote to Jefferson: "My circumstances oblige me to leave, but I take with me the consolation that, while Europe presents an immoral and melancholy spectacle, the people of this continent are advancing with great strides towards the perfection of social conditions. I should like to think that one day I shall again enjoy this consoling experience, and I share your hope . . . that mankind may look forward to great improvements which can be expected from the new order of things to be found here."[2] Of course, Humboldt's last memories of Europe were of the final years of the French Revolution, which ended with the beginning of Napoleon's leadership under the Consulate on December 24, 1799. In the following years, Napoleon sought to establish hegemony over most of continental Europe, which, though ostensibly intended to spread the ideals of the Revolution, in reality led him to consolidate an imperial monarchy that restored some aspects of the Ancien Régime. It is possible that when Humboldt wrote this letter he had already received the information that on April 18, 1804, Napoleon had officially assumed the title of His Imperial Majesty, which would not have given the Prussian cause for optimism. Nevertheless, Humboldt's interest in the United States was apparent even before his visit. In his travel narrative, he mentions that at the end of September 1799, during his stay in Cariaco, Venezuela, he noticed a "marked predilection for the government of the United States" among the people of this region, without mentioning a desire to visit the country himself. It was in this part of the world, he remarks, that he first heard the names "Franklin" and "Washington" pronounced with enthusiasm.[3] It was in Venezuela as well, he comments further, that on December 14 he received the news about the death of George Washington, whose residence Mount Vernon he would visit more than four years later.

Humboldt's View of the New World

In the United States, Humboldt was sure that he had found a model of society that the Spanish colonies in America and European monarchies could, and would, emulate in the future. He followed political and social developments in the new republic attentively for the rest of his life. Outside of the political and social realm, his interest was focused on gold-mining practices (particularly in contrast with the Russian model) and the possibility of building a canal between the Atlantic and the Pacific Oceans, on the basis of which international relations might be improved through free trade. He also appreciated the progress of the United States in the sciences, education, and culture. Nevertheless, Humboldt criticized some aspects of American life in his letters to both European and American correspondents, and sometimes even during personal encounters. Paramount was his opposition to slavery, to which he expressed his strong aversion at every opportunity. He was appalled by the spread of this inhumane institution and by the consequences it threatened for the maintenance of the Union. He also disliked the materialism of American life, the consequences of the California "gold fever," the 1846–48 war between the United States and Mexico, and the expansion of Mormon belief. Humboldt's keen interest in U.S. politics can be attributed to his access to the nation's highest political circles: he maintained contact not only with Jefferson but also with the U.S. presidents John Quincy Adams, James Polk, Zachary Taylor, and Millard Fillmore. During the election campaign of 1856, more than fifty years after his visit to North America, Humboldt went so far as to involve himself in the presidential race, openly stating his preference for John Charles Frémont, the Republican candidate (and an explorer) over the Democrat James Buchanan. Humboldt was a great admirer of Frémont and had included numerous references to his scientific investigations in his *Views of Nature*.[4] He was especially pleased when the explorer based his political campaign on opposition to the spread of slavery into the western territories.

If we are to better understand Humboldt's expectations for the United States as a free society, it helps to take a closer look at his attitude toward colonialism. During his scientific expedition in the Spanish colonies, he experienced the colonial system, with all its injustices for the oppressed or enslaved part of the society, just before the independence movements began to gather force in those territories. In his numerous publications, and even more in his travel diaries, he includes with the scientific information some pointed criticism of the colonial system. One passage from an unpublished essay found in his diaries seems particularly revealing. In it, he condemns the basic concept of colonialism, under which authority was exercised not for the well-being of the inhabitants, but according to the interest of the metropolis. This resulted in a fundamentally immoral system, Humboldt wrote, and created an uneasiness in visitors to the region who were sensitive to these issues. This private document demonstrates that Humboldt's personal convictions were based on strong moral considerations and that the well-being and happiness of a nation's people were always foremost in his mind. Apart from its clear rejection of colonialism as a political and economic institution, this essay stands out for its in-depth analysis of the different facets of colonialism and the possible consequences Humboldt foresees.[5] His consideration of the disparities among the colonial systems established by the European nations demonstrates that his sharp critiques do not refer only to the Spanish colonies he visited: "Nowhere should a European feel more ashamed than on the islands, regardless of whether he is French, English, Danish or Spanish. To debate which nation treats the negroes with more humanity is to mock the word 'humanity,' and to ask oneself whether it would be more humane to slit a person's stomach or to skin them."[6]

Several topics seem to have taken on special relevance for Humboldt during his travels through the Spanish dominions in America, including not only the horrors of slavery, but also the catastrophic consequences of colonialism and the bribery of its administrators, the perilous situation of the Indians, the weaknesses of the missionary system, and the treatment of the workers by the big land or mine owners.

On several occasions, he admitted that he carried the "ideas of 1789 in his heart," always maintaining an awareness of the fundamental principles of freedom, equality, and fraternity upon which his philosophy of life was based. It should be noted, however, that he did not at all approve of the revolutionary fervor and methods of the Jacobins.[7] The terrors of the French Revolution and the cruelties committed during the revolt in Haiti and other slave riots may have influenced his frequent warnings of violent uprisings in America.

That Humboldt undertook his expedition through Spanish America on the eve of its independence makes his description of the prerevolutionary atmosphere in different societies especially interesting. His comments seem surprisingly cautious, however, when considered in light of his strongly expressed feelings about the conditions that were giving rise to an ominous dissatisfaction. While Humboldt made occasional reference to these conditions and the dangers they posed for the rest of colonial America, his diaries offer no insight into his views on the Spanish colonies' claim for liberty. His conviction that the Spanish dominions would find themselves in a more favorable economic position if they achieved their independence from financial interests in Madrid can only be ascertained from his detailed economic analyses, his statistics of the trade, and his comments on the population in different regions, such as in his *Essay on New Spain*.

In part, this reticence may be explained by Humboldt's gratitude to Carlos IV for generously permitting his expedition to pass through the American colonies, the realization of his life's ambition. During the diplomatic preparation for his expedition in the Iberian Peninsula, the undertaking of his scientific exploration of America, and his dissemination of the results of his work, Humboldt was careful to maintain an officially positive attitude toward the Spanish government and its colonial administration in America. His ease of travel, the local colonial authorities' support of his scientific activities, and the likelihood of his being allowed to undertake future expeditions depended on his avoiding critiques of the political situation. Humboldt dedicated his political *Essay on New Spain* to Carlos IV, which he admitted in a letter to Jefferson was intended to soften the attitude of the government in Madrid toward certain individuals in Mexico who had furnished him with more information than the court might have regarded as proper.[8]

Humboldt may also have been moved to prudence by the fate of Alejandro Malaspina, a Spanish naval officer of Italian origin who, after returning to Spain from his expedition to America, Asia, and Australia (1789–94), was accused of spying and imprisoned.[9] When Humboldt left the harbor of La Coruña in northern Spain, sailing toward Tenerife in the Canary Islands, he passed the Castle San Antón, where Malaspina was then being held captive. In his travel narrative, Humboldt comments: "We remained with our eyes fixed on the castle of St. Antony, where the unfortunate Malaspina was then a captive in a state prison. On the point of leaving Europe to visit the countries which this illustrious traveler had visited with so much advantage, I could have wished to have fixed my thoughts on some object less affecting."[10] Later, in his diaries, Humboldt observes that after all the dangers Malaspina had faced and surmounted in his voyages around the world, it was the dangers of Spanish politics that ultimately caused his downfall.[11] Finally, Humboldt himself was first and foremost a scientist, and his cautious attitude was a function of the personal aims he attached to his expedition. Clearly, he was most reluctant to risk his American project.

For these reasons, many of the critical observations on the social structure of the Spanish dominions that Humboldt makes in his *Personal Narrative* fail to appear in his later published works. In some cases, parts of the diaries are even marked "never to be published."[12] And, while Humboldt does give examples in these personal documents of oppression, corruption, and violation of what he considers to be human rights, his writings also often contain proposals for reform in many arenas, from the problem of overreliance on the mono-cultivation of sugar to the elimination of slavery.

As a result of his admiration for the United States and his identification with many of its ideals, Humboldt referred to himself on several occasions as "half an American"[13] or "almost American," as he did in a letter to the German-born Austrian politician and statesman Klemens von Metternich.[14] Humboldt's use of this term permits a double interpretation: not only does he identify with both cultures, but he also holds back from full identification with some aspects of American society, such as slavery, which he regarded as unacceptable in a free nation. As he wrote to the *New York Times* shortly before his death in 1859: "I am half American; that is, my aspirations are all with you; but I don't like the present position of your politics. The influence of Slavery is increasing, I fear. So too is the mistaken view of negro inferiority."[15]

Indeed, one of the most crucial experiences of Humboldt's transatlantic journey was his contact with the system of slavery, which he considered "the greatest of all evils that afflict humanity."[16] He made this clear on many occasions, most notably in his regional study of Cuba[17] as well as his final work, *Cosmos*, in which he asserts his fervent belief in the unity of the human race, rejecting the theory of "superior" and "inferior" peoples. Humboldt believed that there were "nations more susceptible of cultivation, more highly civilized, more ennobled by mental cultivation than others, but none in themselves nobler than others."[18] "We can never enough praise the wisdom of the legislation of the new republics of Spanish America," he writes, "which since their birth, has been seriously occupied with the total extinction of slavery. That vast portion of the earth has, in this respect, an immense advantage over the southern part of the United States, where the whites, during the struggle with England, established liberty for their own profit."[19]

In 1856, John Sidney Thrasher, a pro-slavery southerner, published his English translation of Humboldt's essay on Cuba under the title *The Island of Cuba by Alexander von Humboldt*, conveniently omitting the seventh chapter, in which the author treated the institution of slavery in great detail. Discovering this, Humboldt protested immediately and angrily in a letter to the *Spenersche Zeitung* in Berlin. In this document—translated and published in, among other newspapers, the *New York Times*, *New York Tribune*, and *New York Herald*[20]—Humboldt proclaims this part of his work to be of "greater importance than . . . any astronomical observations, experiments of magnetic intensity, or statistical statement." He was, he writes, "entitled to demand that in the Free States of the continent of America, people should be allowed to read what has been permitted to circulate from the first year of its appearance in a Spanish translation." This assertion drew immediate attacks on Humboldt from the angry defenders of slavery in the United States, and won approval from the abolitionists.

Humboldt's bitter remarks to his friend Karl August Varnhagen von Ense about Buchanan's victory in the U.S. presidential election of 1856 are therefore unsurprising, since for Humboldt this represented the victory of the slavery: "And that shameful party which sells Negro children of fifty pounds, that distributes canes of honor just as the Russian Tsar distributes swords of honor . . . , which demands that free workers should rather be slaves than free men, has won. What a crime!"[21]

Because the institution of slavery conflicted with his idealized image of the United States as a model liberal society, Humboldt hoped—to some degree was even convinced—that slavery would disappear. The United States, he felt, was constantly improving its material circumstances and developing dynamically toward a better society for all its inhabitants, in contrast to Europe, where, Humboldt complained, social and political development was stagnating or even regressing. In 1825, he observes to the geographer Heinrich Berghaus that the issue of slavery had the potential to tear the nation apart: "Should the question of slavery break out one day, I entirely share your opinion that the maintenance of the North American Union as a state is in danger. I do not wish to see this happen. I think highly, very highly of the United States because it is the shelter for a reasonable freedom."[22]

It is apparent in Humboldt's extensive correspondence on the topic that as he gradually realized that the practice of slavery was not vanishing but in fact expanding, he no longer saw any reason to conceal his opinions. In a July 31, 1854, letter to Varnhagen von Ense, he writes scathingly: "In the United States there has, it is true, arisen a great love for me, but the whole there presents to my mind the sad spectacle of liberty reduced to a mere mechanism in the element of utility, exercising little ennobling or elevating influence upon my mind and soul, which, after all, should be the aim of political liberty. Hence there was indifference on the subject of slavery. But the United States are a Cartesian vortex, carrying with them, grading everything to the level of monotony."[23] He had come to believe that the United States no longer embraced the principles of the Age of Reason, and he felt betrayed. The ideal society he yearned for, it seemed, was not to be.

The institution of slavery in the United States was particularly galling for Humboldt because he had hoped that U.S. society would serve as an example for other parts of the world. In the Spanish territories, on the other hand, especially in Cuba, while the horrors of slavery affected him, he considered it as intrinsic to the evils of a colonial society, which would disappear with its independence from the metropolis. It was inconceivable to Humboldt that this inhumane system could persist in a free society and under an enlightened government created in response to the social injustices exhibited in Europe.

One curious example of Humboldt's position on slavery may be found in a Prussian law, passed at Humboldt's instigation on March 9, 1857, stating that the moment an enslaved person stepped into Prussian territory, the master's ownership ceased and that slave was free.[24] Given the rather small number of enslaved persons traveling in Prussia, Humboldt's efforts to establish such a law demonstrate his determination to oppose slavery wherever he had the authority to do so.

Humboldt's interest in the Prussian antislavery legislation should be seen in connection to the U.S. Supreme Court's *Dred Scott* case, which temporally coincided with the emancipation of the Jews in Prussia. In this pivotal decision, the Court ruled in March 1857 that people of African descent, whether or not they were slaves, could never be U.S. citizens, and that Congress had no authority under the Constitution to prohibit slavery in federal territories. The Court ruled further that slaves could not sue in court, and that, being private property, they could not be taken away from their owners without due process. This decision was based on the case of the slave Dred Scott and his wife, Harriet, who had lived in states and territories where slavery was illegal, including Illinois and Wisconsin, which were then part of the Louisiana Purchase. In April 1846, Scott sued for his freedom, arguing that since he had lived in both a free state and a free territory, he had become legally free, but the Court finally ruled that his sojourn outside of Missouri did not affect his emancipation under the Missouri Compromise, since reaching that result would deprive Scott's owner of his property.

Humboldt learned of this case and the final decision of the U.S. Supreme Court through a letter sent to him shortly afterward by the Prussian minister Friedrich von Gerolt.[25] In a letter to John Matthews, Humboldt expresses his indignation about the fugitive slave law: "I have the warmest attachment to your beautiful and liberal city, New York, but have earnestly and deeply regretted that WEBSTER, whom I long respected, more than favored that *shameful* law which still persecuted colored men after they had regained by flight their natural inborn liberty, of which they had been robbed by Christians."[26]

It should be said that Humboldt's portrayal of the North American populace cannot be considered very comprehensive, given the brevity of his visit and that his experience of Americans was largely confined to scientific or political circles. His comments on American society generally portray the United States as the place where his ideals of social progress could eventually be realized.

Jefferson's View of the Old World

Thomas Jefferson's feelings about Europe were as complex as Humboldt's about the New World. The vast number of letters in which he discussed his experiences in Europe, and particularly in prerevolutionary France, illustrate just how profoundly those experiences shaped his ideas about how a new form of society might be built. The letters lay out his beliefs concerning liberty, the structure and obligations of government, and the importance of land ownership for the prosperity of society. His European experience also had a lasting effect on his personal life, his style of living, his artistic and literary tastes and ambitions, and his ideas on virtuous behavior.

When he arrived in France, Jefferson carried with him a fervent belief in republicanism, the virtue of the common man, and the classical ideals of austerity, frugality, and practicality. He was appalled by the French monarchy, which appeared to share none of these values. During his stay, he was alternately shocked by and attracted to

French society. Despite his many negative comments about the social life of the French, Jefferson's account book shows that quite early in his stay in Paris, he began making extravagant and costly purchases in order to keep up with French fashion and customs. These cannot be simply dismissed as the necessary expenditures of a foreign diplomat; Jefferson aspired to be a cultivated and fashionable man of the world. In a lengthy letter written to Charles Bellini, Jefferson offers a detailed portrait of daily life in upper-class, prerevolutionary France, and contrasts the mores of that society with those he saw in the United States.[27] As this communication shows, Jefferson clearly disapproved of the empty daily routine of the European aristocratic class in Paris, filled with what he styled boredom and nonsense. In America, he argues, the days were spent with healthy and useful activity, leisure time was shared with real friends, and there was not only present amusement but the promise of future good. In a letter to John Banister reflecting on what he considered a lack of moral virtue in France, he notes the risk of bad influence and even danger in sending young people to the Old World.[28] An American, he concludes, "coming to Europe for education, loses in his knowledge, in his morals, in his health, in his habits, and in his happiness." Only a few months earlier, he advises James Monroe that traveling to the Old Continent would help him to appreciate America even more: "It will make you adore your own country, its soil, its climate, its equality, liberty, laws, people, and manners."[29] One year later, in a letter to his former mentor George Wythe sent from his post in Paris, Jefferson describes what in his opinion were the sources of the evils in the Old World and concludes: "If anybody thinks that kings, nobles, or priests are good conservators of the public happiness, send them here. It is the best school in the universe to cure him of that folly."[30]

Still, the Old World had much to offer him. At the end of his five years in France, Jefferson returned home with a taste for refined and elegant living. As he wrote to Bellini, "Were I to proceed to tell you how much I enjoy their architecture, sculpture, painting, music, I should want words."[31] Above all he admired European architecture, particularly in its classical forms, which influenced his designs for buildings in his own country. His taste for European art led him to commission the French sculptor Jean-Antoine Houdon to make a bust of George Washington. He held European literature in high esteem, and his fondness for French food and wine was legendary.

Thomas Jefferson's attitude toward slavery is much debated today, in light of both his ambiguous statements about it and his tendency to say one thing and do another.[32] His views evolved over time, as can be seen from his correspondence, his *Notes on the State of Virginia,* and his autobiography. On the one hand, as a figure of the Enlightenment he was a spirited opponent of slavery, which ran contrary to its every tenet. "Under the law of nature," he wrote in April 1770, "all men are born free, every one comes into the world with a right to his own person, which includes the liberty of moving and using it at his own. This is what is called personal liberty, and is given him by the author of nature, because necessary for his own sustenance."[33]

Nor was the "abominable crime," as he called it, tolerable from a religious point of view. "Can the liberties of a nation be thought secure when we have removed their only firm basis, a conviction in the minds of the people that these liberties are of the gift of God?" he asked rhetorically. "That they are not to be violated but with his wrath? Indeed I tremble for my country when I reflect that God is just: that his justice cannot sleep for ever."[34]

Like Humboldt, Jefferson was very aware that the institution of slavery could lead to social revolts or bloody outbreaks of violence. The events in Haiti offered a strong warning, as discussed in chapter 6. Jefferson also knew that slavery tarnished the image of the New World as a better and more just society. While he was convinced that slavery was entirely opposed to the fundamental ideas of the United States and would have to disappear in the long run, he simply did not see how to correct this situation. As he famously summarized the situation: "We have the wolf by the ear, and we can neither hold him, nor safely let him go. Justice is in one scale, and self-preservation in the other."[35]

In *A Summary View of the Rights of British America* from July 1774, Jefferson contends that the "abolition of domestic slavery is the great object of desire in those colonies where it was unhappily introduced in their infant state. But previous to the enfranchisement of the slaves we have, it is necessary to exclude all further importations from Africa."[36] Even at an early stage in his political career he took several steps aimed at bringing slavery to an end: he drafted the Virginia law of 1778, which prohibited the importation of enslaved Africans; in 1784, he proposed an ordinance banning slavery in the new territories of the Northwest; and from the mid-1770s, he expressed support of a plan of gradual emancipation. But after these early efforts on behalf of manumission, Jefferson went through a long period of political silence on the topic. He took no action despite his increasing political power, and he did not prohibit the later expansion of slavery into Missouri and other states. Nor was he among the many Virginia slaveholders who freed their slaves in the period between 1784 and 1806, when the laws of the Commonwealth encouraged manumission.

Jefferson's writings also contain several statements that appear to be supportive of slavery and express a belief in the inferiority of black people, as exemplified by a passage from his *Notes on the State of Virginia* asserting that "blacks, whether originally a distinct race, or made distinct by time and circumstances, are inferior to the whites in the endowments both of body and mind."[37] He viewed black people paternalistically, as children who needed his protection, and was unable to imagine a place for a free African American population within the United States. Worried about a possible "mixture of blood," he viewed the removal of blacks from the country as fundamental if emancipation were to be achieved. He himself was deeply immersed in the system of slavery for his entire life, and owned one of the largest populations of slaves in Virginia. As he recalled it, his first memory was of being carried on a pillow by a mounted slave; and when he lay dying at the end of his long life, it was his trusted slave Burwell Colbert who understood and carried out his master's last wishes.[38]

Jefferson expresses the complexity and difficulty of the matter in one famous passage from his 1821 autobiography. "Nothing is more certainly written in the book of fate, than that these people are to be free," he acknowledges, but then he adds, "nor is it less certain that the two races, equally free, cannot live in the same government. Nature, habit, opinion have drawn indelible lines of distinction between them."[39] For Jefferson, it appears, the peculiar institution was a problem to be solved by future generations.

The question of the extent to which Jefferson's European experiences influenced his beliefs concerning slavery is widely debated. Certainly he consistently manifested a remarkable concern for the socioeconomic circumstances and working conditions of the rural population in any region of Europe through which he traveled. He comments with displeasure in his travel diary "Notes of a Tour into the Southern Parts of France" in March 1787: "I observe women and children carrying heavy loads and laboring with the hoe. This is an unequivocal indication of extreme poverty. Men in a civilized country never expose their wives and children to labor above their forces and sex, as long as their own labor can protect them from it."[40] Unfortunately, he did not find the labor of his women slaves in the United States equally distasteful. In an interesting letter to Thomas Cooper in 1814, he compares the circumstances of American slaves favorably with those of the poor unskilled workers in Europe. Slaves, in his opinion, were

> better fed in these States, warmer clothed, and labor less than the journeymen or day-laborers of England. They have the comfort, too, of numerous families, in the midst of whom they live without want, or the fear of it; a solace which few of the laborers of England possess. They are subject, it is true, to bodily coercion; but are not the hundreds of thousands of British soldiers and seamen subject to the same, without seeing, at the end of their career, when age and accident shall have rendered them unequal to labor, the certainty, which the other has, that he will never want? And has not the British seaman, as much as the African, been reduced to this bondage by force, in flagrant violation of his own consent, and of his natural right in his own person? And with the laborers of England generally, does not the moral coercion of want subject their will as despotically to that of their employer, as the physical constraint does the soldier, the seaman, or the slave? But don't mistake me. I am not advocating slavery. I am not justifying the wrongs we have committed on a foreign people, by the example of another nation committing equal wrongs on their own subjects. . . . I am at present comparing the condition and degree to which oppression has reduced the man of one color, with the condition and degree of suffering to which oppression has reduced the man of another color; equally condemning both.[41]

This long and revealing passage encapsulates Jefferson's views on slavery at a moment quite late in his life, and demonstrates that he had carefully and perhaps defensively considered how the plight of slaves and that of the free working poor of Europe might be measured against each other.

Transatlantic experiences crucially molded the social and political convictions of both Thomas Jefferson and Alexander von Humboldt. The dimensions of the "other" world that most preoccupied each depended on his ideological background and character, as well as his personal interest and position in his own society. Humboldt was most interested in issues connected to his concept of liberty—including personal, political, and economic freedom—and to how that concept related to political institutions such as colonialism and slavery. For Jefferson, Europe provided an opportunity to observe the consequences of a political system that he rejected absolutely. He came away from his time there strongly motivated to create a government that would help the young nation surmount the deficiencies of the Old World and provide happiness to his fellow countrymen.

4 A Transatlantic Network of Knowledge and Ideas

Personal Encounter in Washington

Humboldt's hopes and expectations for his meeting with Jefferson were obvious in his letter of introduction of May 24, 1804. It is a masterpiece of diplomacy, containing every fascinating piece of information Humboldt could think of that might fire Jefferson's enthusiasm as a scientist and politician. Like all his letters to Jefferson, it was written in French, and as always, Jefferson answered in English.

Though in fact he came directly from Cuba, Humboldt wrote in his very first sentence that he had arrived from Mexico, a place he knew to be of great interest to Jefferson. He wished, he said, to personally deliver a parcel to the Virginian from his friend, the U.S. consul in Havana. He further let it be known that he had admired Jefferson and his writings from his own early youth. Despite a "burning desire" to see Paris again, where he expected to publish the fruits of his expedition, he could not resist his "moral interest" in seeing the United States and encountering "a people that understands the precious gift of Liberty."[1]

In a note to Zaccheus Collins, Humboldt expresses the same "moral interest" in becoming acquainted with the United States, a country "that wisely governed."[2] In his letter of introduction to James Madison, he points out his interest in seeing the prosperity of humankind in the United States as a result of its wise legislation. After having seen the majesty of nature in South America, he writes, it is a consoling idea that he will be able to witness the moral spectacle of a free people worthy of this beautiful destiny.[3]

He goes on to emphasize his scientific interest—to be useful to the physical sciences and the study of mankind in its different states of barbarism and culture[4]—and gives a detailed description of his American expedition, which, he notes—highlighting his independence from any European government—has been undertaken at his own expense.[5] Humboldt refers to his works on galvanism[6] and his publications in the Mémoires de l'Institut National de Paris in order to present himself as a prominent scientist with links to the scientific community in Paris. Aware of Jefferson's tenure in Paris fifteen years earlier and knowing that the American president has maintained his scholarly connections there, Humboldt mentions his Parisian contacts, among them Jean-Antoine Chaptal, Louis Nicolas Vauquelin, and, in particular, Georges Cuvier. He also demonstrates his knowledge of Jefferson's *Notes on the State of Virginia,* which he probably had read during his studies in Hamburg at the Commercial Academy with Christoph Daniel Ebeling. One of the most distinguished German experts on the geography and history of America, Ebeling owned numerous works on the United States of which Humboldt made use.[7] A letter from Ebeling to Jefferson offers evidence that the former owned a copy of the 1787 London edition of the *Notes.*[8] Since Jefferson's book makes clear that he knew little of the Indian population of Spanish America, Humboldt, hoping to pique the older man's interest, alludes to that topic as well. In this context, he also mentions his own finding of mammoth teeth in South America at 1,700 toises above sea level, a topic of great interest to Jefferson and one that makes indirect reference to the much-debated degeneracy theory regarding the New Continent.[9] He also discusses details regarding the defense of the United States, a subject Jefferson discusses in *Notes on the State of Virginia.*

In this letter to Jefferson and others, Humboldt attempts to establish a connection between their scientific endeavors by referring to Jefferson as a magistrate-philosopher and scientist rather than as the president of the United States.[10] Also noteworthy is that Humboldt signs the letter to Jefferson with the title "Le Baron de Humboldt" and as a member of the Academy of Sciences in Berlin. Legally, however, Humboldt was not a baron, and at that moment he was an extraordinary member of the Academy of Sciences; he received full membership only after his return to Europe in 1805.[11]

Thanks to this carefully crafted letter that touched upon nearly every Jeffersonian interest and enthusiasm, Humboldt succeeded in drawing the Virginian's attention to both himself and his American expedition, as Jefferson's response, written the day after he received Humboldt's letter, illustrates: "The countries you have visited are of those least known, and most interesting, and a lively desire will be felt generally to receive information you will be able to give. No one will feel it more strongly than myself because no one perhaps views this new world with more partial hopes of its exhibiting an ameliorated state of the human condition."[12] The letter ends with the invitation to Washington for which Humboldt had hoped. Though at that moment, Jefferson wrote, Washington had "nothing curious to attract the observation of a traveler," this paucity of marvels would be more than offset by "the welcome with which we should receive your visit."

It seems peculiar that Jefferson should call the Spanish American countries "the least known" in view of the considerable number of books that had been published on the region, several of which formed part of his own library. In a letter sent in 1787 to the Mexican Miguel Lardizábal y Uribe,[13] who later became a member of the Consejo de Regencia in the court of Cádiz, Jefferson includes a list of the thirteen books on Spanish America that he already owned, written by authors such as Antonio de Ulloa, Francisco López de Gómora, Jorge Juan y Santacilia, Antonio de Solís y Rivadeneyra and Gabriel de Cárdenas y Cano, along with a list of those he wished to acquire. By the time he wrote to Humboldt in 1804, his holdings on Spanish America had very likely increased.[14] In any case, Jefferson's last letter to his colleague, written in 1817, points out that Humboldt's publications on the South American region, "a country hitherto so shamefully unknown," had come at a perfect moment to "guide our understandings in the great political revolution now bringing it into prominence on the stage of the world." While this may be a bit of harmless flattery meant to underscore the importance of Humboldt's writings, it is nevertheless puzzling that he would seem to minimize the value of the many published works on Spanish America. In his reference to the human conditions in the Spanish colonial territories that he would like to see "ameliorated," Jefferson is identifying a topic of common interest and inviting a dialogue that would draw on Humboldt's experience there.

A few days after their first meeting, Jefferson wrote Humboldt a note requesting some practical information on the newly established and still-uncertain frontiers between New Spain and the United States. Specifically, he needed to ascertain whether the western border of the Louisiana Territory was the Sabine River, as the Spanish claimed, or the Rio Grande, which Jefferson called the North River: "We claim to the North river from it's mouth to the source either of it's Eastern or Western branch, thence to the head of Red river & so on. Can the Baron inform me what population may be between these lines, of white, red or black people? And whether any & what mines are within them? The information will be thankfully received."[15] Indeed, the information was significant, since the Louisiana Purchase was the most critical foreign policy problem Jefferson was facing in 1804.

Jefferson and Humboldt's relationship was mutually beneficial. If Jefferson the president was particularly interested in the information about the Spanish colonies that Humboldt possessed, Jefferson the scientist was intrigued by Humboldt the scientific explorer. As president of the American Philosophical Society, Jefferson had for many years been in contact with scientists in Europe and the United States.[16] In his position as president of the United States, on the other hand, he had no formal governmental responsibility for scientific research. Nevertheless, it was his consuming avocation, and over the years, he had created one of the most important scientific libraries in the country. For this reason alone, he would be a valuable professional contact for Humboldt.

As their letters attest, the meeting between Jefferson and Humboldt was largely devoted to a discussion of the border region between New Spain and the United States. For Jefferson, Humboldt translated into French parts of his early work on Mexico, "Tablas geográfico-políticas,"[17] which he had originally prepared for the vice-king of New Spain, José de Iturrigaray. To this he added a two-page commentary on the Mexican border region of the Louisiana Territory, with a special focus on the land between the Río Grande and the Sabine River, in which Jefferson was especially interested. This document, which can be seen as a direct response to the questions Jefferson had raised in his letter of June 9, contains a considerable amount of both descriptive and statistical material touching upon the size, population, climate, and political divisions of various provinces, as well as notes on agriculture and the possible commercial exploitation of minerals.[18]

Jefferson seemed pleased with the information supplied by Humboldt, writing to Caspar Wistar: "I have omitted to state above the extreme satisfaction I have received from Baron Humboldt's communications. The treasures of information which he possesses are inestimable and fill us with impatience for their appearance in print."[19] While Humboldt certainly did not believe that he was transmitting confidential information to Jefferson, his letters provoked a debate over this issue that continues to this day. Some scholars, particularly those who consider the issue from a Mexican perspective, speculate that Humboldt may have been a spy for the Spanish Empire. Some, following the example of Juan A. Ortega y Medina in his introduction to the Mexican edition of Humboldt's *Political Essay on the Kingdom of New Spain*[20]—question the integrity of Humboldt's decision to share information from the Spanish colonial archives with Jefferson. They argue that this intelligence was useful to the United States in its invasion of Mexico forty-three years later, in which the latter country lost a large part of its territory. However, the information Humboldt supplied to the U.S. president in 1804 was much less detailed than that which he published between 1808 and 1811 in his work on New Mexico, information that was available to the public from the time of publication. By the time the United States invaded Mexico in 1847, Humboldt's intelligence would have been well out of date.[21]

Humboldt's attitude toward the exchange of scientific knowledge was a generous one: he felt that it should be free, with no limitations imposed upon it. He had benefited from this open exchange when, shortly before his departure for Europe, he had received recent statistical information on the United States from Albert Gallatin. Humboldt included and duly acknowledged this information in his works *Political Essay on the Kingdom of New Spain* and *Personal Narrative*. Many years later, he returned the favor by helping Gallatin publish an article in the German journal *Hertha*.[22] Humboldt also supplied information on the North American Indians to his brother Wilhelm—information that subsequently appeared in another article in the same volume of *Hertha* in 1827.[23] In his own publications, Humboldt usually acknowledged the source of factual material not only as a courtesy but also as a method of creating a network of traceable scientific knowledge. Thus he was not at all pleased to discover that the explorer Zebulon Pike[24] and the cartographer Aaron Arrowsmith had plagiarized and used for their own publications—without attribution—several maps of New Spain and a report on the region that Humboldt had generously given to Jefferson, to whom he subsequently complained.[25] In a long answer written two years later, Jefferson apologized for what he called the "ungenerous use" of Humboldt's "candid communications." Nevertheless, whereas Jefferson regarded the behavior of Arrowsmith, a Briton, as consonant with "the piratical spirit of his country," in Pike's case he tried to mollify Humboldt by assuring him that Pike, who died as a hero "in the arms of victory gained over the enemies of his country," meant only to enlarge knowledge and not to gain "filthy shillings and pence of which he made none from that book." Jefferson apologized on behalf of the late Pike for failing to acknowledge the source of his information, but maintained it was just "an oversight, and not at all in the spirit of his generous nature."[26]

Unfortunately neither Humboldt nor Jefferson kept an account of their meetings. Nevertheless, they very likely discussed the Lewis and Clark expedition, which after a winter training period in Camp Dubois near present-day Hartford, Illinois, had finally gotten under way on May 14, 1804, just a few days before Humboldt reached the United States. It is unfortunate that Humboldt did not meet the explorers before their departure. After his own five-year expedition through America, he might have been extremely valuable to the explorers with respect to both current factual information and the general experience of such exploration. Apparently, however, Humboldt was unaware of the Lewis and Clark expedition until his meeting with Jefferson in 1804. This is not surprising, since the expedition was originally organized as a U.S. military operation to scout French defenses in Louisiana prior to the purchase of the new territory. By the time of Humboldt's visit, however, the mission was well known, and the two men discussed it freely. Almost a year later, Jefferson was still mentioning details from the conversations with Humboldt on the topic.[27]

Humboldt's scientific interests, of course, ranged well beyond the travels of Lewis and Clark. The diaries of the artist Charles Willson

Peale reveal that he and Humboldt discussed the museum of natural history that Peale was creating, and for which he was trying to obtain government funding:[28] "The Baron came to my room," Peale wrote, "& told me that he had been conversing with the President about me & my Museum, that he wondered that the Government did not secure it by a purchase [of] it—for such opportunities of getting so complete collections of natural subjects seldom occurred. The president replied that it was his ardent wish and he hoped that the period was not far distant & he thought that each of the States would contribute means and thus it might be made a National Museum. The Baron seemed pleased with the subject of their conversation & although he could not relate all that passed yet he assured me that the President was very much my friend."[29]

The enthusiasm that both Jefferson and Humboldt felt for their collegial relationship is clear in their writings of the period. For example, in a letter to Isaac Briggs, written in mid-June during Humboldt's visit, Jefferson relates that he "mentioned to Baron Humboldt my proposition for taking lunar observations at land without using a time piece. He said there could be no doubt of it's exactness, but that it was not new, that even De la Caille had proposed it and Delalande had given all the explanations necessary for it."[30] Here Humboldt offers his American friend a taste of up-to-the-minute research, Nicolas-Louis de la Caille and Joseph-Jérôme Lefrançais de Lalande being among the most important scientists in Paris, then capital of science. For his part, Humboldt wrote in his last letter to Jefferson, sent in 1825, that he had once had the good fortune to discuss with Jefferson, in the Presidential Mansion in Washington, the events that were going to change the face of the world, and that Jefferson wisely had long anticipated.[31]

Just before his departure for France, Humboldt returned to the sentiments he had expressed in his letter of introduction. Regretting that the time had come for him to leave the young nation, he reiterated his vision of the United States and his hopes for the future of Europe. "I have had the good fortune to see the first Magistrate of this great republic living with the simplicity of a philosopher," he wrote admiringly to Jefferson, "and receiving me with that profound kindness that can never be forgotten.[32]

At the time of their meeting in 1804, Jefferson was sixty-one years old and Humboldt only thirty-five. Whereas Jefferson had already attained the highest political position in his country and was a member of the American scientific elite, Humboldt had just concluded the long expedition that would later make him the most famous scientist in the world. Despite his relative youth, he had already built a solid and even illustrious scientific reputation. As an expert in mining he had been retained by the Prussian state; he was the author of several scientific publications, some of which had been translated into other languages; and he had been invited by Baudin to join the team of scientists and explorers who proposed to circumnavigate the globe. His connections to the scientific community of Paris added considerably to his scholarly prestige, and by 1799, his achievements were so impressive that Carlos IV had granted him unrestricted (and unprecedented) permission to explore the Spanish colonies. In view of these attainments, it is obvious that Humboldt was not a mere provider of specific geographic material for Jefferson. The American president also saw him as a person of great interest and an experienced scientist through whom he might forge connections to important European scholars.

Later Correspondence

Over time, Humboldt made certain to keep Jefferson—and through him the New World—informed of the progress of his research and published writings on the topic of his American expedition. At the same time, through his correspondence with Jefferson, he kept abreast of opinions, observations, and ideas from the United States. Jefferson seemed to enjoy sharing his interests and concerns with Humboldt, and he valued the younger man's perspective on these matters. Their personal encounter was only the beginning of a mutually enriching scientific and political dialogue between an ascendant new world and an ever-evolving old one. Humboldt was undoubtedly one of the most energetic correspondents of his day, drafting, it is estimated, an astonishing fifty thousand letters between 1787 and 1859. Jefferson, while considerably less prolific, still managed to write some nineteen thousand letters during his lifetime.[33] Humboldt did not confine his transatlantic correspondence to his famous Virginian friend. After his visit to the United States, he maintained steady contact with Albert Gallatin, James Madison, John C. Frémont, David Bailie Warden, the writer Washington Irving, the scientists Louis Agassiz, Matthew Fontaine Maury, James M. Gilliss, Samuel George Morton, and Benjamin Silliman, the painter George Catlin, the historian George Ticknor, and many others. Among the people Humboldt met during his visit to the United States, Gallatin and Warden appear to have been his most frequent correspondents; he wrote as many as twenty-six letters to each of them. The existing correspondence between Humboldt and Jefferson consists of eight letters from Humboldt and six to him. Since Jefferson used his polygraph to make copies of all his correspondence for his own records, all his letters to Humboldt still exist, but it is likely that some of Humboldt's letters to Jefferson have been lost.[34] Their contact continued almost until the end of Jefferson's life, with the last letter from Humboldt coming to him in 1825.

In these personal documents, they continually express their mutual appreciation of one another as well as of one another's works. In his first letter to Jefferson after his return to Europe, with a renewed awareness of all the distresses of the Old World, Humboldt mentions the admiration he felt for his American correspondent and that he saw him and his ideas as a hope for the future of his own continent.[35] In 1809, another letter to Jefferson is filled with praise of the United States. Humboldt tells his Virginian friend that he has not been happy since he left his beautiful country,[36] and, referring to Jefferson personally, writes: "But what an amazing career yours has been! What a wonderful example you have given of energetic character, of graciousness and depth of the tenderest affections of the soul, of moderation and equanimity as the first magistrate of a powerful nation! What you have created, you see bearing fruit. Your retreat to Monticello is an event the memory of which will live forever in the annals of Mankind."[37] Two years later, Humboldt continued to express his pride in his friendship with the Virginia statesman.[38]

Jefferson responded in kind, expressing "those sentiments of high admiration and esteem, which, altho long silent, have never slept." In another letter, he assures Humboldt that Gallatin would give him, "from time to time, the details of the progress of a country in whose prosperity you are so good as to feel an interest, and in which your name is revered among those of the great worthies of the world."[39] It is clear from the topics discussed that their correspondence extended beyond mere politeness, originating instead from a sincere, mutual interest in an exchange of knowledge and ideas.

The political future of Spanish America, and especially the applicability of democratic models to the region, continued to be of the utmost interest to both men, with each especially eager to know the other's opinion of the independence movement. Their correspondence reveals both jointly held convictions and disagreements on the topic. In an 1811 letter acknowledging the receipt of Humboldt's "valuable works"—the latest publications on his American expedition—Jefferson praises the significance of Humboldt's works, published at a moment "when those countries are beginning to be interesting to the whole world," becoming "the scenes of political revolution" and perhaps eventually "integral members" of the "great family of nations." Preliminaries over, he then expresses his own doubts: "What kind of government will they establish?" he demands skeptically. "How much liberty can they bear without intoxication? Are their chiefs sufficiently enlightened to form a well-guarded government, and their people to watch their chiefs? Have they mind enough to place their domesticated Indians on a footing with the whites? All these questions you can answer better than any other." To some of these questions Jefferson provided a few of his own suppositions. "I imagine they will copy our outlines of confederation and elective government, abolish distinction of ranks, bow the neck to their priests, and persevere in intolerantism," he answered. "Their greatest difficulty will be in construction of their executive. I suspect that, regardless of the experiment of France, and of that of the United States in 1784, they will begin with a directory, and when the unavoidable schisms in that kind of executive shall drive them to something else, their great question will come on whether to substitute an executive elective for years, for life, or an hereditary one. But unless instruction can be spread among them more rapidly than experience promises, despotism may come upon them before they are qualified to save the ground they will have gained."[40]

In his answer to this letter several months later, Humboldt articulates further concerns, though his comments this time are much less detailed. He predicts a violent conflict in Spanish America that would leave its imprint on the social order. Like those that had occurred in European society, it would probably be based on mutual resentments and animosity.[41]

Two years passed before Jefferson replied; because of war with England, mail service to and from the Continent was exceedingly slow. In fact, Humboldt's two letters of December 1811 had come by a circuitous route through a chain of messengers and did not arrive until well over a year after they were written. In this new letter, he outlines the evolution of his thinking and puts forth a rather skeptical view of the long-term practicability of the movement. In this many-faceted analysis, his political predictions and critical reflections on the role of religion are particularly noteworthy:

> History, I believe, furnishes no example of a priest-ridden people maintaining a free civil government. This marks the lowest grade of ignorance, of which their civil as well as religious leaders will always avail themselves for their own purposes. The vicinity of the New Spain to the United States, and their consequent intercourse, may furnish schools for the higher, and example for the lower classes of their citizens.
>
> And Mexico, where we learn from you that men of science are not wanting, may revolutionize itself under better auspices than the Southern provinces. These last, I fear, must end in military despotisms. The different cast of their inhabitants, their mutual hatreds and jealousies, their profound ignorance and bigotry, will be played off by cunning leaders, and each be made the instrument of enslaving others. But of all this you can best judge, for in truth we have little knowledge of them to be dependent on, but through you.[42]

A few years later, in 1817, when it was easier to see which direction the events in the former Spanish colonies were taking, Jefferson reiterates his skepticism about their fitness for self-government: "The issue of it's struggles, as they respect Spain, is no longer a matter of doubt. As it respects their own liberty, peace & happiness we cannot be quite so certain. Whether the blinds of bigotry, the shackles of the priesthood, and the fascinating glare of rank and wealth give fair play to the common sense of the mass of their people, so far as to qualify them for self government, is what we do not know."[43]

Some of the comments Jefferson made in his early years could lead to the assumption that he would support liberal democratic revolutions in general, and that he saw the American Revolution as a model for independence movements in other nations. This was not the case. Jefferson was frequently critical of other independence movements; he did not view the American case as exportable, and repeatedly argued that the most successful ingredient of any reformist or revolutionary movement was time. In concert with the political thinker Montesquieu, Jefferson believed that state traditions and histories could be successful only if they changed gradually over time; and that attempts to impose new modes and orders on a culture with no experience of them would inevitably fail, possibly creating a worse state of social affairs. If the fruit on a tree needed violent shaking to come loose, he observed metaphorically, it was a signal that the fruit wasn't mature enough to stand alone. If it were fully ripe, however, the fruit would fall naturally from the branch.[44]

Having witnessed the beginnings of the French Revolution during his time in France, Jefferson had developed a complex attitude toward revolutionary movements—a construction of his ideals, personal interests, political convictions, and firsthand experiences in the American Revolution. This attitude shifted over the years, partially as a result of developments in France, but also through his own evolving political philosophy. He argued that France had no experience with liberal republican government and that the autocratic power of monarchy and church were too ingrained in the culture to be terminated by a simple declaration of the rights of man. Accordingly, he offered Parisian reformers rather conservative advice, recommending that they negotiate by small increments with the king, who would gradually transfer power to an assembly while retaining enough royal privilege to prevent a counterrevolution. As early as 1791, Jefferson anticipated much violence in France, and the rise of military despotism later realized under Napoleon.

His analysis of the independence movements in the Spanish-speaking parts of the Western Hemisphere was identical to his perspective on the events in France, and followed a similar logic. In a candid and revealing correspondence with John Adams between 1817 and 1822, the two men analyzed how these matters affected U.S. interests in the southern part of America. Jefferson agreed with both cause and objectives of the independence movements, and hoped that the movement's leaders would achieve their goals, but he did not believe they could. The result, he predicted, would be military tyrannies. Though the revolutions might be successful in the short term, Jefferson was skeptical about their consolidation into workable democratic states. Without the established organs of an independent legislature and judiciary, freedom of the press, and secular law, he believed, society would soon revert to prerevolutionary patterns, just as Napoleon had revived some of the Bourbons' authoritarian traditions and centralized power. In Jefferson's view, once Spain was expelled, the real problems would begin, as they had in France after the execution of Louis XVI.[45] In an 1818 letter to Adams, he comments: "I do believe it would be better for them to obtain freedom by degrees only, because that would by degrees bring on light and information and qualify them to take charge of themselves understandingly, with more certainty if in the meantime under so much control only as may keep them at peace with one another. Surely it is our duty to wish them independence and self-government, because they wish it themselves, and they have the right and we none to choose for themselves."[46]

In his last letter to Humboldt, written in 1817, Jefferson elaborates on this view. "The first principle of republicanism," he observes, "is that the lex majoris partis is the fundamental law of every society of individuals of equal rights: to consider the will of the society enounced by the majority of a single vote as sacred as if unanimous, is the first of all lessons in importance, yet the last which is thoroughly learnt. this law once disregarded, no other remains but that of force, which ends necessarily in military despotism. this has been the history of the French revolution; and I wish the understanding of our Southern brethren may be sufficiently enlarged and firm to see that their fate depends on it's sacred observance."[47]

Surprisingly, however, Jefferson seemed to have no interest in establishing a dialogue with the protagonists of the Spanish American independence movement, as he did, for instance, with the revolutionary Marquis de Lafayette or the writer, economist, and government official Pierre Samuel du Pont de Nemours. The more progressive Humboldt had already developed relationships with the Venezuelans Andrés Bello and Simón de Bolívar, the Ecuadorean Vicente Rocafuerte, and the Mexican Lucas Alamán.[48] Jefferson, however—in contrast to the position he adopted during the French Revolution—maintained a certain distance both from the revolutionary leaders of Spanish America and the events in which they were central figures.

Jefferson was well aware of Humboldt's political influence among intellectuals in the French capital, which may explain the American president's extensive speculations on the future of Latin America and on the role of the United States in its political and social development. On these topics, he solicited Humboldt's opinions, but the Prussian, as usual, was circumspect in his replies. Jefferson, on the other hand, seems more willing in his later correspondence with Humboldt to openly express his political opinions than he showed himself to be in an earlier letter, in which he remarks, "On politics I will say nothing, because I would not implicate you by addressing to you the republican ideas of America, deemed horrible heresies by the royalism of Europe."[49]

In the midst of the war between the United States and England, and years before the Monroe Doctrine, Jefferson briefly summarized for Humboldt the relationship between America and Europe, and the emerging idea of American isolationism. "The European nations constitute a separate division of the globe; their localities make them part of a distinct system; they have a set of interests of their own in which it is our business never to engage ourselves," he contends. "America has a hemisphere to itself. It must have its separate system of interests, which must not be subordinated to those of Europe."[50]

Humboldt, who did not share this conviction, which conflicted directly with his vision of an interconnected world, remained silent on the issue. For him, international trade and the peaceful transfer of ideas and goods remained a high priority. This exchange illustrates well Jefferson and Humboldt's differing priorities: while Jefferson focused on a nation's political and economic independence, Humboldt was most concerned with the progress of science.

Jefferson continued this letter, written in 1813, touching upon another delicate subject: U.S. policies regarding the indigenous population. These he was eager to present in a positive light:[51]

> You know, my friend, the benevolent plan we are pursuing here for the happiness of the Aboriginal inhabitants in our vicinities. We spared

> nothing to keep them at peace with one another. To teach them agriculture and the rudiments of the most necessary arts, and to encourage industry by establishing among them separate property. In this way they would have been enabled to subsist and multiply on a moderate scale of landed possession. They would have mixed their blood with ours, and been amalgamated and identified with us within no distant period of time. On the commencement of our present war, we pressed on them the observance of peace and neutrality, but the interested and unprincipled policy of England has defeated all our labours for the salvation of these unfortunate people.

Humboldt, in his response, chose to focus on the 1809 selection of Madison as Jefferson's successor. He offers congratulations and mentions his very positive impression of the new president. He agrees with Jefferson that this new development "promises a wise and honest administration."[52]

Given the fact that Humboldt addressed the plight of the native population in Spanish America in several of his publications,[53] the lack of any reply to Jefferson's comments on the North American Indians seems a strange omission. The fact that none of Humboldt's reflections on U.S. Indian policies can be found in his texts might be explained by his limited experience of North American society. Nevertheless, one may wonder why he did not use the correspondence with Jefferson to complete his picture of American native peoples.

The Lewis and Clark expedition continued to be a topic of mutual interest to the two men even into the late years of their correspondence. Jefferson was aware of Humboldt's intense curiosity about the explorers' discoveries and kept him informed him on the impending publication of the expedition notebooks. Inconveniently, this came in the middle of the Anglo-American War of 1812–15. "You will find it inconceivable that Lewis's journey to the Pacific should not yet have appeared, nor is it in my power to tell you the reason," he writes. "The measures taken by his surviving companion Clark for the publication, have not answered our wishes in point of dispatch. I think however, from what I have heard, that the main journal will be out within a few weeks in 2 vols. 8°. These I will take care to send you with the tobacco seed you desired, if it will be possible to escape the thousand ships of our enemies spread over the ocean."[54]

Their letters were filled with astronomical observations as well, and discussions of ambitious plans for a canal between the Pacific and Atlantic Oceans. This was one of Humboldt's pet projects, and armed with his considerable geographic expertise, he often proselytized about the advantages of a possible interoceanic connection. Additionally, providing support and letters of recommendation to younger travelers and scholars for the furtherance of scientific knowledge was of significant interest to both men.[55] Only one major topic of the day is missing from the Humboldt-Jefferson correspondence: that of slavery. Given the fact that slavery was then being hotly debated in transatlantic dialogues, it seems strange that two such illustrious public intellectuals would fail to address the issue in theirs. What role did it play in their relationship, given their distinctly different personal and political backgrounds? No record of any discussion remains. Given Humboldt's essentially diplomatic character, his finesse in dealing with Spanish authorities, and his initial overture to Jefferson, he likely realized that broaching such a controversial issue with the president of the United States might make for an uncomfortable encounter, and prove contrary to his own interests in the long run. Perhaps if they had met at Monticello, it would have been more difficult to ignore the slavery question.

Nevertheless, in communication with Jefferson, Humboldt alluded occasionally to comments on slavery that he had inserted in his work on New Spain. In a letter written in May 1808, he enclosed the first part of this work, mentioning to Jefferson that he would find his name cited with the enthusiasm it has always inspired in the friends of mankind.[56]

In the passage cited, Humboldt discusses the U.S. slave population, which had increased considerably after the Louisiana Purchase, and contends, incorrectly, that at that time the slave trade had been interdicted. His confusion probably stemmed from the fact that the Constitution had banned the importation of slaves, a ban that took effect in 1808. Humboldt maintains that the ban on importation would have happened long before if the law had permitted Jefferson (whom he calls a "magistrate whose name is dear to the true friends of humanity") to oppose the introduction of slaves.[57] Here

Humboldt adds a footnote to introduce Jefferson to readers as the "author of the excellent Essay on Virginia," without mentioning that at the time of the Louisiana Purchase, Jefferson was still the U.S. president. Interestingly, Humboldt defines Jefferson as an author—thus making his work on Virginia known to a larger public—and presents him as a person acting against the institution of slavery rather than as a slaveholder himself. No doubt Jefferson appreciated this courtesy, which elevated his scholarly credentials while avoiding a discussion of his personal connection to slavery.

This letter underscores two points. First, Humboldt continued to be very interested in and well informed about the legal circumstances of slavery in the United States, and was particularly aware of Jefferson's position with regard to it. Second, Humboldt seemed determined to make public only those aspects of Jefferson's actions that suggested support for abolition. For whatever reason, he was intent upon conveying an image of the American president's convictions that mirrored his own. Obviously, since Humboldt planned to return to the United States at some point and hoped to remain on good terms with Jefferson and his other American friends in the southern elite, he avoided provoking controversy.

Humboldt, ever the scientist, wanted to ensure that his works would contain the most recent U.S. legislation on American slavery, but, tellingly, he sought confirmation from the diplomat David Bailie Warden rather than Jefferson.[58] Humboldt asked whether he had committed an error supposing "that the importation of slaves into the united states is not totally interdicted, and also whether there be any facts or observations, concerning the united states, in the Statistical part of his work, which ought to be corrected in the second edition."[59] Jefferson, however, could not be dodged. Warden passed Humboldt's questions along to him, and a few months later Warden received a reply from Jefferson confirming that "the importation of Slaves into the United States is totally & rigorously prohibited."[60]

Only once did Humboldt directly approach the subject of slavery in a letter to Jefferson. In June 1809, he wrote to express anxious regret for his remarks on slavery in the previously mentioned passage of his work on Mexico, in which he had denounced the United States for countenancing the institution. He had learned only after publication, Humboldt writes, that Congress had taken serious steps toward "total abolition." He had been carried away by his devotion to the cause of enslaved Africans, he adds, and he now felt ashamed of his comment. He promises to repair the error in future editions with a special note and an appendix at the end of his work. These apologetic and rather poignant remarks demonstrate how strongly Humboldt wished to view Jefferson as an ally in the fight against slavery. Since the United States did not move toward abolition in the following years, however, Humboldt let his criticism stand without modification.

Jefferson remained silent on this subject, and during the remaining sixteen years of their correspondence neither he nor Humboldt brought up slavery again—this despite the fact that the debate continued throughout both of their lifetimes and had a significant place in many of their writings. With his other American correspondents, Humboldt handled this issue differently. While still in the United States, he addressed the topic in a letter to the abolitionist William Thornton. Having read Thornton's *Political Economy: Founded in Justice and Humanity* (1804), Humboldt considered him to be like-minded on the topic, and thus wrote more emotionally and unguardedly than he would have in his correspondence with Jefferson. In this letter, Humboldt points to the level of liberty in the United States, stating that in this country the small bad things were compensated for by the considerable goods.[61] The legislation permitting the importation of slaves to South Carolina, he declares further, was abominable and disgraceful, particularly in a country where many enlightened people lived. Humboldt recognized that an interdiction of this inhumane practice might initially lead to a reduction in cotton production, but he did not accept this argument, since he detested the politics that measured and evaluated the public welfare simply according to the value of its exports. He saw the wealth of nations in terms of the well-being of individuals and concluded his letter to Thornton with the spirited proclamation that before one was free, one had to be just, and that without justice there was no lasting prosperity.[62]

But if slavery was off-limits as a topic of discussion for Jefferson and Humboldt, countless other subjects were not. Another subject of common interest, though not discussed in their letters, was their formulation of a response to the ideas of the Comte de Buffon, William Robertson, Cornelius de Pauw, and Guillaume-Thomas Raynal, rejecting their theory of the inferiority of American species.

Both showed an appreciation of the French scientist and his professional achievements in general, but sought to refute his work in this particular debate. In fact, Jefferson was the first person in the United States to formulate a written response to the Buffon's ideas.[63] His analysis was detailed: In his *Notes on the State of Virginia,* particularly in Query VI, he contrasts the size of particular animals of the New World with the same species in the Old, and uses his findings—along with others concerning the existence of certain plants and the nature of the climate—to provide scientific evidence challenging Buffon's theory. Humboldt, whose enthusiasm for America encompassed both the natural world and its human inhabitants, declares himself in firm opposition to Buffon's assumptions as well. Even before his American expedition, he was familiar with those arguments, later calling them "unphilosophical" and contrary to generally acknowledged physical laws. In numerous references to Buffon, Robertson, and others, he reveals what he considered to be a lack of knowledge about their subject or a lack of liability of their sources, pointing out their mistakes in detail. Humboldt asserts, for instance, that Robertson's *History of America* "only values the revenue of Mexico at four millions of piastres, while it actually amounted at that period to more than eighteen millions. Such was the state of ignorance in Europe at that time respecting the colonies of Spain that that learned and illustrious historian when treating of the finances of Peru, was compelled to derive his information from a manuscript drawn up in 1614."[64] Humboldt had still more harsh words for Robertson, saying that he, Raynal, and de Pauw "disfigure the names of cities and provinces."[65] Of Buffon, he offers a more general critique, noting that it was "superfluous to refute here the rash assertion of M. de Buffon, as to the pretended degeneracy of the domestic animals introduced into the New Continent. These ideas were easily propagated, because, while they flattered the vanity of Europeans, they were also connected with brilliant hypotheses, relative to the ancient state of our planet."[66]

On several occasions, he proves these writers' conclusions to be wrong, "destitute of truth," or "unskilled," such as Raynal's assertion that domestic animals transported to Portobello become infertile.[67] In his work *Views of the Cordilleras,* he makes reference to the ideas of some unnamed distinguished writers who, "more struck with the contrasts than the harmony of nature, have described the whole of America as a marshy country, unfavourable to the increase of animals, and newly inhabited by hordes as savage as the people of the South Sea. In the historical researches respecting the Americans, candid examination had given place to absolute skepticism."[68] Humboldt proclaims that he preferred the American climate to all others because "there one could breathe more freely."[69] Here he refers literally to the climate, in response to claims about its inferiority, but also figuratively to the country's status as a free nation: he notes his belief that many Europeans exaggerated the influence of the American climate on the spirit of the people, asserting the impossibility of intellectual work being undertaken productively there.[70] Humboldt and his travel companion Bonpland were living proof of the opposite: they swore they had never felt more powerful than they did as they contemplated the lavish natural beauty of the New World. Its greatness electrified them, filled them with joy and made them have the sensation of invulnerability.[71]

Debates over the imputed "inferiority" of the New World were active on both sides of the Atlantic. Besides Jefferson and Humboldt, many others participated in their own publications, though in most cases they did not garner the attention accorded the two internationally known Enlightenment figures. Obviously, Americans felt challenged by Buffon's and de Pauw's theories of the degeneracy of life forms in the New World.

At this time, South American scholars were starting to publish their own works, thus making their arguments and scientific data accessible to a larger public. Among the first to publish defenses of America were several Jesuits, the avant-garde of the Creole culture, though most of their refutations were limited to discussions of their own region.[72] One of them, Francisco Javier Clavijero Echegaray, notes in the preface of his work on the history of Mexico (*Storia antica del Messico*) that he published his work with the aim of restoring the ancient splendor of his native country, which had been obscured by Buffon, de Pauw, and Raynal. He refers to these writers by name, pointing out what he considers to be their errors and contradictions and contending that their works on America were not based on personal experience and lacked reliable information. Besides employing irony and counterattack as polemical tactics, he also reverses the Europeans' arguments, turning their own reasoning back on them and thus illuminating the Eurocentric character of the degeneracy debate.

A strong part of Clavijero Echegaray's refutation was the defense of the American Indian. According to him, the existing dissimilarities with the Europeans were not natural, but social—meaning, for example, that indigenous Americans would need better education in order to equal Europeans in science.[73] Juan Ignacio Molina, in his work on Chile (*Compendio della storia geografica, naturale, e civile del regno del Chile*), lovingly details the flora, fauna, and geography of his native country, and defends it with "nostalgic yearning for the beauty of the Chilean landscape and the mildness of its climate," aiming to convince his reader of the benevolence of Chilean nature.[74] He sharply criticizes the writings of de Pauw and Buffon, but while he excoriates de Pauw's ignorance, he expresses respect for the Buffon's authority, even though he does not agree with him. His defense is based on the proposition that American nature is different from but not inferior to that of European.[75] Juan de Velasco's work *Historia moderna del Reyno de Quito* comes to the defense of the kingdom of Quito, in an effort to protect it against those "anti-American philosophers" who would "reduce the glory of his native country."

Other New World opponents of the American degeneracy theory produced more subtle works of resistance. Their method was to marshal a corpus of scientific arguments and data that opposed degeneracy theories while underscoring the many marvels of American nature. José Manuel Dávalos can be listed among these. His work concentrates on the climate of Peru, which he maintains was particularly healthy, with any physical problems of its inhabitants caused by other factors such as food and improper eating habits. Similarly, the Chilean Manuel de Salas claims for his nation a privileged climate, perfect for human happiness as well as for the production of all the animals and plants of Europe, and yet without wild beasts, troublesome insects, or poisonous reptiles.[76] He also contrasts the youthfulness of the New World with old and tired Europe. In *Observaciones sobre el clima de Lima*, published in 1806, José Hipólito Unanue y Pavón asserts that it was not climate, but moral factors, such as slavery, that made men weak and idle.[77] An important contributor to the climate debate was Francisco José de Caldas, who from 1808 published the scientific journal *Semanario del Nuevo Reino de Granada.*[78] Caldas considers that his native country offered excellent possibilities to observe "the influence of climate and diet on the physical constitution of man, his character, his virtues, and his vices." This assertion led to an interesting discussion in the *Semanario* with his compatriot Diego Martín Tanco: Tanco states that climate had no impact on either the moral or the physical growth of human beings because everything depended on men's beliefs and upbringing. Caldas counters that a certain connection existed between the environment and man's physical development as well as his moral character, but that humans were still free to choose between good and evil.[79] According to these writers, climate did not determine America's inferiority to Europe.

Many Creole replies to climate theories were based on intense study and personal observation. The controversy over the impact of climate on living beings—humans in particular—was a particular preoccupation that inflected many other fields: botany, medicine, geography, and—more potently and dangerously—race theory, and therefore colonialism. The American interest in refuting the inferiority theory shows an obvious desire to present the continent in a better light and the struggle of some for political independence. It shows another central Enlightenment concern as well: the control of the natural world by men. This is where the distinction between civilization and wilderness, order and chaos is truly established.[80]

Particularly in America, with its grand and unexplored nature, there was an intense desire to dominate nature as a sign of civilization. Consequently, the notion that humans were dominated or determined by nature was variously received by Americans: some accepted the idea of the climate's influence on the character and the physical constitution of men, although they had diverse opinions regarding the extent of this process. Others attributed the difficult circumstances in which large parts of the American population had to live exclusively to the impact of colonial oppression, lack of education, and poverty rather than to environmental conditions.

The "degeneracy" debate was crucial to the natural sciences of the period and produced a large number of scholarly works on both sides of the Atlantic. Numerous scientific writers waded into the fray, testing their own ideas in argument and thus giving considerable impetus to the study of nature and the progress of science. By provoking this transatlantic exchange, the degeneracy debate actually rendered a service to eighteenth-century intellectual life. The fact that experience and empirical knowledge were used to refute the theory showed that it was becoming difficult to maintain "scientific" theories based on irrational beliefs and unreliable or incomplete data.

It might seem surprising that, in the context of his argument against the degeneracy theory in his *Notes on the State of Virginia,* Jefferson did not establish a connection to those people who professed the same view in Spanish America. This can be partially explained by the fact that he limited his study to Virginia. Nonetheless, his refutation of Buffon was in a larger sense also a defense of the "new" American continent, an aim he shared with those Spanish American authors. Another likely explanation for the absence of reference to these Creole scholars in Jefferson's defense of his home country is the early date of publication of the *Notes* (1782).[81] Nevertheless, few comments on these publications are to be found in his later writings. From one of the few existing references, we do know that Jefferson not only possessed Clavijero's work on Mexico, *Storia antica del Messico,* but that he was aware of its argument. He notes in a letter to Joseph Willard in 1789 that "Clavigero, an Italian also,[82] who has resided thirty five years in Mexico, has given us a history of that country, which certainly merits more respect than any other work on the same subject. he corrects many errors of Dr. Robertson, and the sound philosophy will disapprove many of his ideas. we must still consider it as an useful work, and assuredly the best we possess on the same subject."[83] By 1806, Francisco de Miranda had finally sent Jefferson a copy of Molina's *Compendio della storia geografica* (commissioned sixteen years earlier, as he wrote in a letter to William Short in 1790).[84] A letter to Charles Willson Peale reveals that he also appreciated "Molina's account of Chili," in which the author corrected "Buffon's classification of the wooly animals."[85] Taking into account the enormous importance that the defense of the New Continent had for Jefferson, his lack of connection to the Spanish American authors who were also engaged in that defense may suggest that he did not see the refutation of the European theory of American inferiority as a joint American venture.

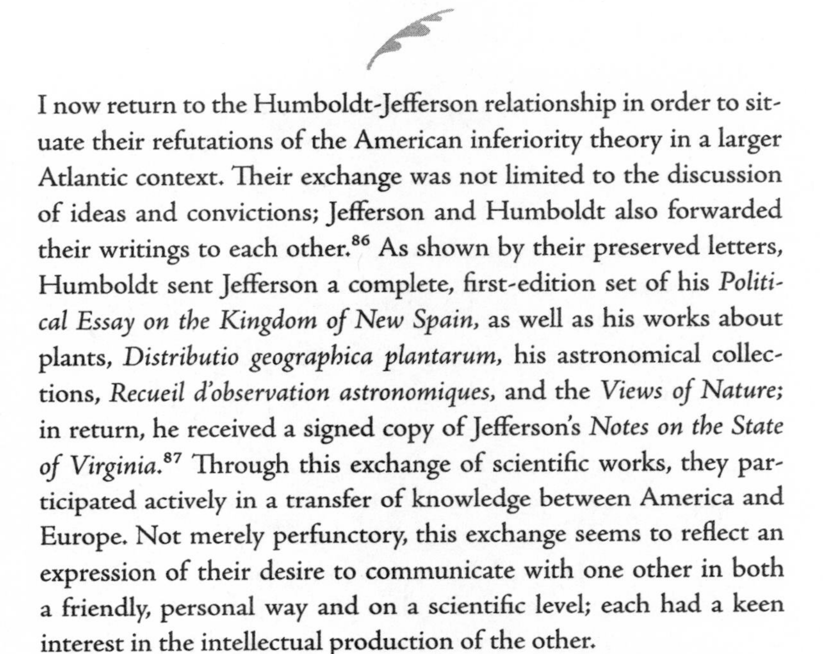

I now return to the Humboldt-Jefferson relationship in order to situate their refutations of the American inferiority theory in a larger Atlantic context. Their exchange was not limited to the discussion of ideas and convictions; Jefferson and Humboldt also forwarded their writings to each other.[86] As shown by their preserved letters, Humboldt sent Jefferson a complete, first-edition set of his *Political Essay on the Kingdom of New Spain*, as well as his works about plants, *Distributio geographica plantarum*, his astronomical collections, *Recueil d'observation astronomiques*, and the *Views of Nature*; in return, he received a signed copy of Jefferson's *Notes on the State of Virginia*.[87] Through this exchange of scientific works, they participated actively in a transfer of knowledge between America and Europe. Not merely perfunctory, this exchange seems to reflect an expression of their desire to communicate with one other in both a friendly, personal way and on a scientific level; each had a keen interest in the intellectual production of the other.

At that time, Jefferson's *Notes on the State of Virginia* was already recognized as a significant contribution to early American literary, an important document on the politics and society of Revolutionary America.[88] Though Humboldt had owned this book for many years, in a letter in 1809 he expressed his desire to receive a signed copy: "I possess your excellent work on Virginia, but should like to receive it from your own hands, with a line of your handwriting. It would be a very precious memento. You made me a gift of your own copy of Playfair, but your name is not in it, and I am afraid of this public misery, which divides everything into lines of red or blue." Referring to a common friend in Paris, he coaxed gently: "Please do not decline my request. Madame de Tessé, who is as devoted to you as I, says that my request is quite reasonable."[89] Nevertheless, more than a year later he had to ask again: "I entreat you to make me a gift of your work on Virginia."[90] After two more years, he finally received the prize he sought, as he confirmed immediately in a letter written the next day.[91]

Given the years separating their publications—Jefferson wrote his only book more than twenty years before he met Humboldt, whereas the latter published most of his books after returning from his American expedition—a useful comparison of their references to each other in their respective publications is not possible. Nevertheless, in later years Jefferson added by hand in his own copy of the *Notes on the State of Virginia* information that Humboldt had sent—for instance, the existence of perpetual snow,[92] and he referred to Humboldt's first formal description of the leaves of the *Espeletia frailejon* in 1801.[93]

Humboldt's publications, however, are full of allusions to Jefferson. Apart from the previously mentioned reference in his work on Mexico to the issue of slavery, there are also comments on Jefferson amid praise for Lewis and Clark's efforts: "This wonderful journey of Captain Lewis was undertaken under the auspices of M. Jefferson, who by this important service rendered to science has added new claims on the gratitude of the savants of all nations."[94] In this publication, his *Political Essay on the Kingdom of New Spain*, he alludes to Jefferson's research on the subject of the American languages, and on other occasions Humboldt also refers to the refutation of Buffon's inferiority hypothesis in "the excellent work of Jefferson." He points to particular pages in Jefferson's *Notes on the State of Virginia*, proving that he knew the book well; he cites Jefferson in connection with Raynal. When he finds it necessary, he even corrects his Virginian friend: "Mr. Jefferson, in his classical work on Virginia, has discussed the position of the Presidio de S. Fe in New Mexico; he believes it to lie in 38° 10′ of latitude; but striking a medium between the direct observations of M. Lafora and Fathers Velez and Escaiante, we shall find 36° 12′."[95] In his *Personal Narrative*, Humboldt mentions Jefferson's publication on Virginia as well as his description of the *Megalonyx*.[96] Finally, he again cites Jefferson and his *Notes* in his own *Views of the Cordilleras*.[97] Most of these references appear in an intellectual or scientific context, presenting Jefferson as an author, scientist, or magistrate.

For both Humboldt and Jefferson, identity was clearly a product of contrasts between the New World and the Old, and each maintained a vivid, lifelong interest in the political and social events of their respective "other worlds." They understood that the happenings in their own milieux were intrinsically connected to the developments on the other side of the Atlantic. And perhaps most importantly, both Humboldt and Jefferson hoped to see their utopian visions realized not only in a national context but in that of the world.

What did this relationship of long standing mean to the two men? Certainly it was not devoid of self-interest. For Jefferson, it offered contact with one of the great specialists of Latin America living in Europe, a friend to some of the most illustrious scholars in Paris, and a member of prestigious learned societies. For Humboldt, the mere fact of his access to the President's House measurably increased his prestige and influence in the French capital. At the time of their meeting, Humboldt was interesting to Jefferson primarily for his trove of information about Mexico. Years later, when Humboldt's scientific prestige had substantially grown, Jefferson's demeanor toward the much younger man changed considerably. In a letter accompanying a copy of his *Notes on the State of Virginia,* he remarks modestly, "They must appear chetif enough to the author of the great work on South America."[98] The fact that Jefferson was twenty-six years older than Humboldt undoubtedly contributed to this evolution of his view, since the Prussian's fame and high international recognition was still to come in the fifty-five years after the finalization of his American expedition.

Many of Jefferson's correspondents, particularly the Europeans, knew about his friendship with and interest in Humboldt, and passed along news of the Prussian's scientific activities and publications. For instance, in 1809, Madame de Tessé sent a letter to Jefferson announcing that she wanted to send him an engraving of Humboldt—an "illustrious traveller, a devotee of your government, and a great admirer of your person." She did not "doubt in the least that Mr. Humboldt will be very flattered to find himself at Monticello when he learns of it, but I have, however, whatever my nephew might say about it, less a desire to please him by sending it than pleasure in repaying him."[99] She wrote a few months later that this "very true-to-life engraved portrait of Mr. Humboldt that I thought would be a lasting reminder of me in your office" was lost,[100] but she was able to find another one, which she was going to send through Count Theodore Pahlen, the Russian minister plenipotentiary to the United States. Two years later he finally received this engraving, accompanied by a letter from Pahlen explaining the delay and telling him that his friend in Paris knew about "the interest you take, Sir, in this traveling scholar."[101] Also in 1809, John Vaughan informed Jefferson that Humboldt's publications of his American expedition were starting to appear and added that Humboldt had expressed his desire to find an American bookseller interested in purchasing the rights for the English edition of his works.[102] One day later, Vaughan wrote again, this time to convey the news that he had just received Humboldt's work *Nivellement barometrique*.[103] Jefferson answered only a few days later, commenting that "Baron Humboldt's work is voluminous & expensive, but it will add much new & valuable information to several branches of science. I have received one part of it and have some others on their way. one part has unfortunately miscarried, & is that which I should have valued most, on the geography of plants."[104] A letter from Lafayette apprised him of the fact that at that moment "our friend Humboldt" was preparing for an expedition to Bengal and Tibet.[105] A few months later, Jefferson received a letter from David Bailie Warden, who told him that Humboldt had not received the note forwarded to him by John Adams, and at that moment was in Germany. Shortly he would return to Paris and hoped to hear soon from his Virginian friend.[106]

This information indicates that many people around Jefferson knew of his closeness to Humboldt and were aware of his interest in the progress of Humboldt's scientific work and possible plans to undertake other expeditions. Even years later, Jefferson continued to mention Humboldt, which inspired his correspondents to continue to keep him informed on the Prussian's activities. When Jefferson's first biographer, Henry S. Randall, wrote to Humboldt asking about his visit to the president, he mentioned that several members of Jefferson's family "have testified to me that Mr. Jefferson often spoke of you and always with that great respect which his public correspondence exhibits. They testify, too, that he always mentioned you with particular personal kindness. They often heard their mother (Mr. Jefferson's oldest daughter, Mrs. Randolph, now deceased) state the same facts and speak of you as having been a guest of Mr. Jefferson's at Washington."[107]

It is revealing that in their letters the two men referred to each other in the third person, and in a rather flattering way. Each seemed pleased and impressed by his correspondent both professionally and personally, and both were undoubtedly gratified by the friendship. Jefferson calls Humboldt on several occasions simply "the baron," which shows that he apparently was impressed by the title Humboldt used, and praises his vast knowledge; Humboldt refers to Jefferson as magistrate, philosopher, friend of humanity, and, in a letter to William Thornton, describes both Jefferson and Madison as "moral phenomena" who "leave[] a kindly impression in one's soul."[108] At the end of his life, writing to to Alexis de Tocqueville, Humboldt calls Jefferson "l'homme ilustre" and mentions that they stayed in contact many years after his visit to the United States.[109] Since Jefferson also liked to portray himself as a philosopher and "amateur politician," we can assume that he very much appreciated such characterizations. Finally, each man seemed to be proud of his important and very prestigious connection to the Old and New World respectively.

Without doubt, the correspondence between these two Enlightenment thinkers was also influenced by Humboldt's increasing significance in North America, not only in the scientific community—where many declared themselves Humboldt's disciples—but also among writers, artists, explorers, educators, politicians, and the general public. In the fields of literature and the arts, Ralph Waldo Emerson, Henry David Thoreau, Edgar Allan Poe, Washington Irving, Walt Whitman, Julia Ward Howe, Oliver Wendell Holmes, and William Prescott all found inspiration in the German naturalist's work,[110] and they readily expressed their admiration and intellectual debt to their "hero of knowledge."[111] Thoreau followed Humboldt's model of plant ecology and at the end of his life was still analyzing the Prussian's impact on environmental thinking in his country.[112] Among American geographers Humboldt was influential as well.[113] The landscape painters of the Hudson River school, Frederic Edwin Church foremost among them, responded to Humboldt's call for the integration of careful observation with the aesthetic response to nature.[114] Church traveled to South America to see the places and peoples Humboldt had described in his works, producing his famous *Chimborazo* (1864) and other paintings.

In the second half of the nineteenth century, many academics attested to Humboldt's influence on American education. Arnold Guyot, a Swiss émigré who brought the study of geography to the

United States, acknowledged Humboldt's overarching importance to the discipline. Louis Agassiz, the Harvard University zoologist generally regarded as the founder of natural science in the United States, was a Humboldt protégé, and Humboldt's accounts of his journey through Spanish America inspired a generation of explorers who were setting out to survey the western United States, among them John Frémont, Charles Wilkes, Amiel W. Whipple, and Alexander Dallas Bache. This demonstrates Humboldt's considerable contribution to the progress of science in general and the impact of what is called "Humboldtian science" also in the United States.

The publication of his final work, *Cosmos*, added substantially to Humboldt's fame in this country and elsewhere, and drew the attention of the general public as well that of scientists and intellectuals. Its first translations into English appeared between 1845 and 1848, shortly after publication of the original German version. In 1855, the Philadelphia publisher F. W. Thomas prepared a German edition of the first three volumes of this work for German-speaking Americans, to which he added the fourth volume in 1869.[115] The 1838 inauguration of steamboat service markedly shortened the voyage across the Atlantic and made the European experience accessible to an increasing number of people. Subsequently, visiting Humboldt in Berlin became an excursion highlight for many a traveling American.[116]

By the time Humboldt died on May 6, 1859, he had attained wide recognition and respect in the United States.[117] The centennial of the birth of the great scholar-explorer—September 14, 1869—was commemorated in cities across America with parades, speeches, concerts, banquets, and the dedication of monuments.[118] Subsequent references to him in the popular press throughout the remainder of the nineteenth century attest to his enduring popularity. In the United States, more places were named after him than in any other place in the world: his name was given to towns, streets, and counties across the nation; lakes were named after him, as were a river, a bay, a marsh, and a reservoir; there were Humboldt Flats and Humboldt Heights; a Humboldt Mountain, Hill, Peak, Sink, and Forest; and numerous Humboldt Parks, as well a Humboldt Cave and a Humboldt Mine.[119] As America expanded and new territories were settled, Humboldt's name was affixed to sites and features from Pennsylvania to California, from Texas to Canada. A group of Free State immigrants to Kansas trumpeted their antislavery politics by naming their town after the Prussian,[120] and the state of Nevada would have become the state of Humboldt if the proposal had been approved at the Constitutional Convention of 1864.

As the then–secretary of war John Buchanan Floyd wrote Humboldt in 1858: "Never can we forget the services you have rendered not only to us but to all the World. The name of Humboldt is not only a household word throughout our immense country, from the shores of the Atlantic to the Waters of the Pacific, but we have honoured ourselves by its use in many parts of our territory, so that posterity will find it everywhere linked with the names of Washington, Jefferson and Franklin."[121]

All this can be seen on a broader scale as part of the international Humboldtian network, which benefited both Humboldt and the persons referring to him. On one side, some Americans used the famous Prussian scientist as an international reference in order to lend their works an international status and make them known abroad; on the other side, Humboldt used his contact to important personalities in North America in the context of his own scientific work as well as to support political, social, or technological causes that had remained important to him through all these years.

So how can the contrast be explained between Humboldt's position as a popular hero in the nineteenth century and the fact that he and his accomplishments have only recently been rediscovered in the United States, after languishing in relative obscurity throughout the twentieth century?[122] Several factors suggest themselves: The publication of Charles Darwin's *On the Origin of the Species* in 1859—the same year Humboldt died—with its great impact in the natural sciences and the subsequent shift of the scientific paradigm, gradually eclipsed the Prussian's fame. Darwin's deterministic "survival of the fittest" credo, which saw the natural world as a battleground, contrasted sharply with Humboldt's romantic approach to nature, which was predicated on the idea that natural diversity would create order and harmony, an idea that, after Darwin, seemed somewhat old-fashioned and naive.[123] With the further progress of science and the invention of modern scientific instruments capable of much more precise measurements, some of Humboldt's results became obsolete. His conviction that the interconnections between all scientific branches ought to be investigated became unpopular with the rise of scientific specialization and the birth of modern academic disciplines in the mid-nineteenth century. Focus on the American Civil War and its disastrous consequences diminished enthusiasm for Humboldt and his discoveries, as did the slow rise of anti-German sentiment in the United States during the late nineteenth century, a disenchantment that culminated in World War I.[124] Humboldt had always been particularly revered in the German community in the United States, but the descendants of these immigrants were much more concerned with assimilation than with honoring their cultural heritage.

Given that the United States was a new nation in search of cultural reference points, it is not surprising that Jefferson's influence in Europe does not compare with Humboldt's in America. Nevertheless, it is singular that Europeans perceived Jefferson not only as a Founding Father and president of the United States, but also as a scholar and recipient of scientific honors in Europe. In 1797, he was elected as a member of the Board of Agriculture in London and of the Linnean Society of Paris; in 1809, he was honored with membership in the Dutch Royal Institute of Sciences, of Literature and of Fine Arts; and in 1814, he was granted membership in the Agronomic Society of Bavaria.[125] These honors bestowed by European scientific institutions suggest the international impact of Jefferson's intellectual work.

5 Jefferson Presents His New Nation

The creation of a new form of society in America fundamentally different from that of Europe was a subject that preoccupied Thomas Jefferson throughout his life. In his early years, he saw the American Revolution as the beginning of fundamental political and societal changes that could spread through the Old World and Latin America, and he envisioned the role the United States might play as the leader of such a worldwide movement. Later, however, he became aware of the limited applicability of the U.S. model. During his years in Paris, Jefferson was the principal political intermediary between France and the United States, and thus the person best positioned to transmit American thought to Europe and to facilitate the exchange of ideas between the two countries.

From the earliest days of the new republic, Jefferson recognized the importance of promoting the American idea in the Old World. In his extensive correspondence with Alexander von Humboldt and many other Europeans—most of them members of the political, scientific, and intellectual elite—he disseminated his personal visions and convictions regarding America in general and the United States in particular, and thus presented a specific image of this young nation.

What image of the American republic did Jefferson seek to disseminate among Europeans, and on which aspects of it did he focus?[1] It is interesting to examine how Jefferson modified this image over time, and the ways in which this process was shaped by both external changes in the political and social situation of the United States and by the changing conditions in France following the French Revolution, as well as the evolution of his own perception of events taking place on both sides of the Atlantic. It is also revealing to see how Jefferson's description of his country changed depending on his correspondent.

In certain circles in Europe, there was enormous interest in the United States and in what was called "the American experiment"—the construction of a new social and political system after the overthrow of British colonial rule. Many were eager to obtain any information about the New World, and they worked to acquire it through European travelers and American citizens abroad. Thomas Jefferson held a key position on both sides of the Atlantic for providing such information. On the one hand, he created an image of Europe for his American correspondents, and on the other, his letters to his European friends helped to shape their idea of the new United States.

Jefferson worked assiduously, with vigorous arguments and long lists of natural-historical evidence, to refute the theories of American inferiority promoted by Buffon and his associates. As an American foreign minister, Jefferson was charged with promoting economic growth and negotiating commercial treaties, and thus with presenting his country in a way that would encourage immigration and commercial relations with other nations. He also had to counter the negative image of the United States created by the British press. He was keenly aware of what was published about America in the widely read French *Encyclopédie méthodique*, so when, in 1786, Jean-Nicolas Démeunier, the editor of *Économie politique et diplomatique*, one of the dictionaries constituting the *Encyclopédie*, asked for his advice on drafts of articles on the United States and a number of its individual states, Jefferson gladly took the opportunity to serve as a consultant and contributor.[2]

Jefferson saw his nation as a model for others to follow, and he therefore—and not only in the interests of his country—made himself the center of a busy import-export network for both ideas and information, of which his correspondence formed an essential part. He was an extremely active participant in what he called the "republic of letters" that bridged the worlds of enlightened men, and he was fully aware of the importance of these letters in the creation of national images, both at that historical moment and for future generations. He also knew that his writings circulated secretly among European liberals. Jefferson's active engagement in molding the image of the United States in Europe made his relationship with

Alexander von Humboldt, whose fame and influence was growing in the European scientific and intellectual world, particularly significant for him from the time of their personal meeting until the last years of their correspondence. In an effort to make Humboldt an advocate for his causes, Jefferson wrote to Humboldt about most of the issues of concern to him. An examination of the image of his new nation that Jefferson sought to promote to Humboldt provides important information into their long-lasting relationship.

The many Europeans with whom Jefferson exchanged letters included revolutionaries, diplomats, scientists, philosophers, members of the military, economists, merchants, and writers. He also corresponded with a number of women, some of whom, particularly those in Paris, were related to men involved in the independence movement and therefore called themselves "américaines." The majority of them, such as Madame de Tessé, Madame de Bréhan, and Madame de Corny, were great admirers of the American experiment who sought to introduce into France not only the political institutions of the United States, but also the democratic habits with which they had become acquainted.[3] Jefferson was aware of the fact that these women and the salons they organized or frequented represented an influential and potentially useful force. Important issues were often discussed in these social gatherings, and some of these women no doubt exercised influence over their husbands or male relatives.

Even at the beginning of the American Revolution, Jefferson was endeavoring through his private communications to influence European opinion on America. In a 1775 letter to John Randolph, he expressed the fear that falsehoods about the United States were being disseminated abroad. Jefferson was responding, in particular, to the idea circulating that American opposition to British rule was confined to a small faction of malcontents and that most Americans were cowards who would readily surrender. Jefferson was determined that Europeans should become "thoroughly and minutely acquainted with every circumstance relative to America as it exists in truth. I am persuaded this would go far towards disposing them to reconciliation."[4]

He further alluded to the difficult situation of a young nation surviving without trained armed forces and plagued by enormous financial problems. In 1781, during the Revolutionary War, he confided to Lafayette in two letters his worries about the military weakness and inexperience of the United States,[5] and a few years later he admitted to the French author Marquis de Chastellux that these military confrontations had caused "the total destruction of our commerce, devastation of our country, and absence of the precious metals."[6] Subsequently he reacted badly to Chastellux's 1781 travel narrative *Voyage de Newport, Philadelphia, Albany &c*, which contained remarks Jefferson considered unfavorable to the United States. Accordingly he presented Chastellux with a lengthy proposal suggesting how the book might be modified before it was translated into English, including the deletion of certain passages and specific instructions for correcting what Jefferson saw as its many obvious errors.[7]

In 1785, he wrote another letter to Chastellux in which he revealed his two great objectives for Virginia: the emancipation of slaves and the establishment of the Constitution on a more permanent foundation.[8] He also made it clear that he was very much involved in contesting the "degrading" theories of French naturalists, one of his major concerns of the moment.

Among the numerous letters Jefferson wrote to his female correspondents in Europe, the most important is probably one addressed to the Anglo-Italian artist and composer Maria Cosway. It contains the famous "head versus heart" dialogue that delineates the two sides of his personality: intellectual and scientific on one side; imaginative and passionate on the other.[9] He offers a highly romantic description of nature: "The Falling spring, the Cascade of Niagara, the Passage of the Potowmac through the Blue mountains, the Natural bridge. It is worth a voiage across the Atlantic to see these objects; much more to paint, and make them, and thereby ourselves, known to all ages. And our dear Monticello, where has nature spread so rich a mantle under the eye? Mountains, forests, rocks, rivers. With what majesty do we ride above all storms! How sublime to look down into the workhouse of nature, to see her clouds, hail, snow, rain, thunder, all fabricated at our feet! And the glorious Sun, when rising as if out of a distant water, just gilding the tops of the mountains, and giving life to all nature!"[10]

He cannot resist alluding to the aspersions cast by the British press, writing contemptuously: "When you consider the character which is given of our country by the lying newspapers of London, and their credulous copyers in other countries; when you reflect that all Europe is made to believe we are lawless banditti, in a state of absolute anarchy, cutting one another's throats, and plundering without distinction, how can you expect that any reasonable creature would venture among us?" "But you and I know," he continues, "that all this is false: that there is not a country on earth where there is greater tranquility, where the laws are milder, or better obeyed: where every one is more attentive to his own business, or meddles less with that of others: where strangers are better received, more hospitably treated, and with a more sacred respect."

Finally, he focuses on the process of nation building. Americans, he writes, are occupied in "opening rivers, digging navigable canals, making roads, building public schools, establishing academies, erecting busts and statues to our great men, protecting religious freedom, abolishing sanguinary punishments, reforming and improving our laws in general."[11]

Jefferson took advantage of the outbreak of the French Revolution to present his country as an ideal refuge in turbulent times. In another letter to Cosway, he declares his intention to leave the scenes of tumult in Paris, where all was politics and even love had lost its part in conversation. He was returning to a country, he adds, where love was not a "consolatory thing" but in its "sublimest degree," whereas in great cities, love was "distracted by the variety of objects," and friendship suffered from the same cause.[12] A year later, when Jefferson was back in New York, he sent Cosway a letter drawing a seductive picture of life in the United States: "You make children there, but this is the country to transplant them to. There is no comparison between the sum of happiness enjoyed here and there. All the distractions of your great cities are but feathers in the scale against the domestic enjoiement and rural occupations, and the neighborly societies we live amidst here."[13] In other notes he attempted to lure Cosway to the United States "to draw the Natural bridge, the Peaks of Otter &c." and to visit with him "all those grand scenes."[14]

No doubt the enticing image of the United States that Jefferson crafted through his letters to Maria Cosway also reflected his personal interest in seeing her again, but a similar—if more restrained and less emotional—character imbued his letters to other women friends. To Madame de Bréhan, for instance, he sent a romantic description of the American wilderness, encouraging her to "go and visit the magnificent scenes which nature has formed upon the Hudson, and to make them known to Europe by your pencil."[15] Likewise, Madame de Tessé is invited to enjoy the "genial climate, a grateful soil, gardens planted by nature, liberty, safety, tranquillity and a very secure and profitable revenue from whatever property we possess."[16]

In the aggregate, these letters to women mostly concern themselves with sublime nature, and often encouraged the recipient to travel to the United States to see for herself and then describe or paint the New World in order to make it better known to Europeans. Mindful of his audience, he stressed his country's peacefulness, happiness, and domestic pleasures.[17] This romantic and idyllic idea of the nature of his country can also be understood in the context of his arguments against the assumed inferiority of America.

In general, as American minister in Paris, Jefferson stressed the advantages of the United States over Europe; and during and after the French Revolution, he presented his home country as a peaceful place, to which he invited several of his European correspondents. Nevertheless, to those he most trusted, Jefferson also revealed the problems the young nation was facing, as well as the political tendencies of which he disapproved. In a letter to the French philosopher and politician Comte de Volney, for example, he complained that the citizens were "divided into two political sects. One which fears the people most, the other the government."[18] Among others who had gained Jefferson's confidence were Lafayette and the English revolutionary Thomas Paine.[19] To his Italian friend Philip Mazzei he wrote quite confidentially until 1796, when Mazzei made public without Jefferson's knowledge a letter containing comments highly critical of American foreign policy, which was then translated and published in French and Italian newspapers.[20] One paragraph in particular was interpreted as a personal criticism of George Washington, thus making Jefferson the target of intense Federalist attacks. Although he had written to Mazzei as a private citizen, the incident and its unpleasant aftermath dogged Jefferson for many years and made him a much more cautious correspondent.[21] To the Polish revolutionary Andrzej Tadeusz Kósciuszko[22] he mentions that on politics he "must write sparingly, lest it should fall into the hands of persons who do not love either you or me."[23] Indeed, in his next letter to Mazzei, Jefferson tells him quite frankly that his transatlantic correspondence from then on would be of a private nature, since "the practise of publishing intercepted letters for political purposes, has prevented my writing a line to any body on the other side of the Atlantic."[24]

Jefferson portrayed Americans as a sovereign people able to appoint political institutions to exercise their authority; they were citizens of a peaceful nation that had been forced into war.[25] He wanted the United States to be a "model for the protection of man in a state of *freedom* and *order*," and he hoped its citizens would be able to "preserve here an asylum where your love of liberty & disinterested patriotism will be for ever protected and honoured."[26] This perspective informed his ideas for the creation of a university in Virginia: writing in 1800 to Pierre Samuel du Pont de Nemours, he expresses a wish to omit those branches of science "no longer useful or valued[27] . . . and introduc[e] others adapted to the real uses of life and the present state of things."[28]

In letters to his European correspondents during his years as president, Jefferson persisted in dwelling on the special features of the American model and the American people. In 1801, he opined to Thomas Paine that the United States should not get involved with the powers of the Old Continent, even in support of shared principles. European interests, he felt, were so different from those of the United States that they must "avoid being entangled in them."[29] A year later, in a note to the British natural philosopher Joseph Priestley, Jefferson enumerated the national characteristics that made possible the successful creation of a new society in America. He considered the people to be wise because they were under the "unrestrained and unperverted operation of their own understanding." The American nation, he writes, "furnishes hopeful implements for the interesting experiment of self-government," adding that "circumstances denied to others but indulged to us, have imposed on us the duty of proving what is the degree of freedom and self-government in which a society may venture to leave its individual members."[30]

He emphasized the personal safety that American citizens enjoyed, as well as the importance of the freedom of the press. The latter was of special significance to him, and its realization a point of pride.

Margaret Bayard Smith made note of a revealing incident that occurred during Humboldt's visit in the spring of 1804.[31] In order to emphasize the freedom of the press permitted in his country, one day Jefferson showed Humboldt some press clippings containing severe personal and political criticisms of himself. According to Smith, Humboldt asked Jefferson: "Why are these libels allowed? Why is not this libellous journal suppressed, or its Editor at least, fined and imprisoned?" Jefferson encouraged him to take the clippings back with him and show them in Europe: "Put that paper in your pocket Baron, and should you hear the reality of our liberty, the freedom of our press, questioned, show this paper, and tell where you found it." Smith relates that Humboldt liked to tell this and similar anecdotes to demonstrate why he so admired Jefferson.

At this point in his life, the Virginian statesman looked upon the young United States with satisfaction after the long years of nation building. He frequently noted the nation's advances in the military sector and the sciences, and the improvements in its economy. In an 1809 letter to Du Pont, he mentions that the special circumstances of his country at this early stage had generated an enthusiasm for home manufacturing. Each household could now fabricate the principal items, reducing to a minimum the number of articles for which they were dependent on others. He also points out the great progress Americans had made in the art of printing, so that they no longer had to import all their books from England.[32] In a letter to Kósciuszko some months later, he focuses on the military buildup, describing U.S. armaments "superior to any we have ever seen from Europe."[33] In the same letter, he makes an intriguing allusion to self-censorship. Formerly, since he could not be sure his correspondence was absolutely confidential, he had to avoid political comments, which made his letters "necessarily dry." But now, "retired now from public concerns, totally unconnected with them, and avoiding all curiosities about what is done or intended, what I say is from myself only, the workings of my own mind, imputable to nobody else."[34]

Another important matter for Jefferson in the early years of his retirement was the political situation in the Spanish part of America, and the first signs of the independence movements in these regions. He mentions the topic to several correspondents in Europe, expressing both ideas and fears for the future of the colonial societies, and, as usual, pointing to the disparities between the United States and any other place—in this case, Spanish America. This topic figures prominently in his letters to Humboldt, since Jefferson considered him an expert, having been an eyewitness during his scientific expedition through the territories. Jefferson foresaw difficulties ahead for the Spanish American colonies on their way to independence: religious intolerance, the construction of a political executive, despotism—all were dangers unless the populace received direction from outside. He writes in a similarly pessimistic vein to Kósciuszko[35] and Du Pont, telling the latter that he feared "the degrading ignorance into which their priests and kings have sunk them, has disqualified them from the maintenance or even knowledge of their rights, and that much blood may be shed for little improvement in their condition. Should their new rulers honestly lay their shoulders to remove the great obstacles of ignorance, and press the remedies of education and information, they will still be in jeopardy until another generation comes into place, and what may happen in the interval cannot be predicted, nor shall you or I live to see it."[36]

In happy contrast, the United States continued its upward trajectory, at least as Jefferson saw it, based on the growth and prosperity achieved during the previous thirty years in both civic life and military institutions.[37] Regarding military strength, Jefferson believed that for the moment America had nothing to fear from its enemies, since it was now able to respond to all kinds of attacks in kind. As to civil life, he pointed to the rapid expansion of manufactures, which he considered nearly on the footing with those of England, going so far as to say that Britain did not have "a single improvement which we do not possess."[38] Merino sheep, he noted, were spreading across the continent, leading to the production of fine cloth "equal to the best English."[39] His last letter to Humboldt portrays the United States as a marvel of technological innovation and offers his own stirring forecast for the future:

> In our America, we are turning to public improvements. Schools, roads and canals are every where either in operation or contemplation. The most gigantic undertaking yet proposed is that of New York for

drawing the waters of Lake Erie into the Hudson. . . . The expence will be great; but it's effect incalculably powerful in favor of the Atlantic states. Internal navigation by steam boats is rapidly spreading thro all our states, and that by sails and oars will ere long be looked back to as among the curiosities of antiquity. We count much too on it's efficacy in harbor defense; and will soon be tried for navigation by sea. We consider this employment of the contributions which our citizens can spare, after feeding, and clothing, and lodging themselves comfortably, as a more useful, more moral, and even more splendid, than that preferred by Europe, of destroying human life, labor and happiness.[40]

Jefferson's frank criticism of the Old World can be seen as a response to the complaints in Humboldt's previous letters about the devastation in Europe caused by political upheavals such as the Napoleonic Wars, which had started with the First Italian Campaign in 1796 and extended until the War of Seventh Coalition in 1815.

There is a passage in Humboldt's travel narrative that can be understood as a reference to Jefferson's reflections here, as well as to a previous message from December 6, 1813. He refers to the progress occurring in the New World and his own rather positive vision of the future relationship between the two continents: "The population of the New Continent yet surpasses but little that of France or Germany. It doubles in the United States in twenty-three or twenty five years; and at Mexico, even under the government of the mother country, it doubles in forty or forty-five years. Without indulging too flattering hopes of the future, it may be admitted, that in less than a century and a half the population of America will equal that of Europe. This noble rivalship in civilization, and the arts of industry and commerce, far from impoverishing the ancient continent, which has been so often prognosticated, at the expense of the new, will augment the wants of the consumer, the mass of productive labour, and the activity of exchange."[41]

Though Jefferson tended to contrast his home nation with England, occasionally he turned to France. As usual, Europe came up short. In an 1816 letter to Pierre Samuel du Pont, he calls the citizens of the United States "constitutionally and conscientiously democrats" who "consider society as one of the natural wants with which man has been created."[42] He goes on to say that both nations "consider the people as our children, and love them with parental affection.

But you love them as infants whom you are afraid to trust without nurses; and I as adults whom I freely leave to self-government."

To a few trusted correspondents, however, he also mentions the problems he foresaw for times to come. To his long-standing friend and confidant Lafayette, he articulates worries about slavery, and America's interest in bringing it to an end: "All know that permitting the slaves of the south to spread into the west will not add one being to that unfortunate condition, that it will increase the happiness of those existing, and by spreading them over larger surface, will dilute the evil everywhere, and facilitate the means of getting finally rid of it, an event more anxiously wished by those on whom it presses than by the noisy pretenders to exclusive humanity."[43] In another letter to Lafayette, written in 1823, he comments on the conflicts of interest between the North and the South. Regarding the forthcoming elections,[44] he returns to the issue of slavery, which in his opinion was being used as a tool for electioneering purposes in the northern states.

In 1823, the Greek humanistic scholar Adamantios Korais wrote to ask Jefferson for guidance in drawing up a suitable constitution for a Greece liberated from the Ottoman Empire. Jefferson responded with a long, considerate letter offering numerous suggestions. Here he emphasizes that the only legitimate objects of government chosen by the people were the equal rights of man and the happiness of every individual. He underscores the relevance of public education and points out the importance of certain principles for the protection of life, liberty, property, and the safety of the citizen—such as freedom of religion, freedom of person, trial by jury, and legislation and taxation only by the representatives of the people. He concludes by stating that, in particular, the freedom of the press should be assured, "for it was the best instrument for enlightening the mind of man, and improving him as a rational, moral, and social being."[45]

What becomes clear from his correspondence is that Thomas Jefferson promoted the New World in Europe through various political, military, economic, scientific or sentimental images, with the topics discussed dependent on the interests and position of each correspondent and the level of trust existing between Jefferson and that person. This last element was a very important one, particularly after the publication of the letter to Mazzei and its unpleasant consequences. While he nevertheless continued to send some political news across the Atlantic while he was still in office, Jefferson spoke out most openly on controversial issues after his retirement. As long as he held political position, he seemed reticent to discuss certain topics, such as slavery, which was too controversial, and religion, which he may have considered too private.

Interestingly, Jefferson's correspondence promotes an image not only of the political and social structure of the United States, but also of America as a continent, in all its diverse components: climate, topography, flora, fauna, and so forth. This suggests the important question of Jefferson's own level of personal identification: What did Virginia, the United States, and the American continent mean to him? From the comments in the letters discussed here, one might conclude that he saw himself as a Virginian first, and the United States as the abstract political system, still under construction, to which his native country, Virginia, belonged. The American continent, in Jefferson's letters, appears as the natural environment in which this project was being realized, offering the necessary freedom and resources, as well as endless and beautiful countryside.

Undoubtedly, the events that most strongly impacted Jefferson's understanding of the United States were the French Revolution and the subsequent revolts and independence movements on the one hand; and the political, social, and economic progress of his country on the other. Although Jefferson initially was optimistic about the changes in France that led to the French Revolution, he later became more aware of the uniqueness of the American political experiment and began to extol it in his letters, pointing out the particular circumstances of the United States that seemed to predict its success. In the letters written while he was living in France, Jefferson admitted how advanced Europe was compared to his home country in regard to the arts, architecture, and culture more generally. This sentiment would later change, and he increasingly voiced his conviction that the United States was no longer trailing behind culturally. There was nothing left to be assimilated from Europe.

During his entire life, Jefferson defended the reputation of his homeland—its climate, its natural phenomena, its government, its people—and he used his official as well as his private correspondence to do so. Because of his position and his own symbolic capital,[46] his information and ideas were well received in Europe. His letters provide valuable insights into the birth and development of a nation, including the diverse difficulties and obstacles it had to face. Finally, the letters offer numerous details of the transatlantic relationship and the history of intellectual exchange between the Old and the New World in the last part of the eighteenth and the early part of the nineteenth century.

As we have seen, Alexander von Humboldt was of considerable use to Jefferson in disseminating an image of America that Jefferson crafted for European consumption. Jefferson was well aware of Humboldt's contact and correspondence with the most prominent personalities in Europe, his scientific publications, his popular book directed to the general public, and the numerous artistic representations of the American continent with which he illustrated this book. Conscious of the Prussian's sympathy for the United States in general and the American project in particular, Jefferson may have seen Humboldt as an ally in his attempt to put enlightened ideas into practice in establishing a new form of society in the New World. He discussed many issues of importance to him in their correspondence and always expressed interest in Humboldt's opinions. The fact that Humboldt had himself experienced the conditions of social oppression in colonial Spanish America and had personally lived under several European monarchies characterized by the social injustices Jefferson sought to overcome made him an interesting interlocutor. Humboldt's letters, filled with enthusiasm for the United States, contributed decisively to his appeal for Jefferson, since in them he transmitted his personal reflections on the New World to the Old. His experience of Europe, colonial Spanish America, and the United States was singular at the time and contributed substantially to the authority Humboldt was accorded by his contemporaries. This, besides the two men's personal affinity, was a very important aspect of their correspondence. Humboldt was well aware of his role in the United States as its champion and promoter in the Old World, which is made evident in a letter to Albert Gallatin. Acknowledging receipt of statistical data Gallatin had sent him, Humboldt promises to tell the world about the altogether admirable and benevolent financial administration of the new country.[47] In the next decades, as Gallatin continued to send Humboldt the latest statistical information for use in his works, Humboldt kept his promise to disseminate his positive view of the commercial success of the United States on a global scale, thus fulfilling the expectations Jefferson had had for him from the beginning.

6

Two Views of the Haitian Revolution

The slave insurrection on Saint-Domingue began in August 1791 under the leadership of the former domestic slave Toussaint Louverture. It reached its first victory when slaves defeated the French colonial forces in 1801 and culminated in a proclamation of independence and the foundation of Haiti as a free republic by Jean-Jacques Dessalines in January 1804. Haiti—the name adopted by the country after independence was declared, restoring the name used by the island's original settlers—became the first country in the Western Hemisphere to eliminate the institution of slavery. The Haitian Revolution, not just the most successful rebellion directed by a black and enslaved population ever to have occurred in the Americas but also the only one that led to the founding of a state, was a defining moment in the history of Africans in the New World. Haiti can be regarded as an ideological laboratory for Enlightenment actors, in which they tested how far their principles of liberty, equality, fraternity, and the right to revolt against oppression extended. The revolution provoked immediate reaction in both the New and Old Worlds, where responses were tightly linked to personal interests and convictions and to the political system the observers inhabited.[1] The slave population of the New World saw an example of what might be obtained, and white society perceived the possible result of the brutal colonial system they had established and expanded in the seventeenth century. The principal consequence was that whites became fearful and determined to prevent similar developments in the region—for instance, through instituting a more humane, more efficient, and overall more stable slave system.

The revolt in the French colony Saint-Domingue attracted a great deal of attention since it involved several popular controversies of the period: colonialism as a European institution expanded over the rest of the world; the independence movements that developed as a reaction; the system of slavery; the aims and the results of the French Revolution; as well as the realization of the humanistic postulates of the eighteenth century. For Enlightenment intellectuals, the Haitian Revolution seemed a test of their belief in the principles of liberty and equality for all people, and in the people's right to revolt against the abuse of power. Humboldt and Jefferson held differing views of the Haitian Revolution. In Jefferson's case, he also had the political power to influence events on the island.

As discussed earlier, Humboldt vigorously rejected colonialism, along with slavery, and repeatedly warned of the possibility of violent revolt. Jefferson's response displayed more ambivalence. He was now president of the United States—the first independent republic in the hemisphere—in the crucial moment of the revolution as well as in the first years after the 1804 Haitian declaration of the independence.[2] In his early years, Jefferson was convinced of the probability of future revolutions not only in the colonial regions of America, but also in the absolutist monarchies in Europe. Regarding the tumultuous political situations of Holland and France, he wrote to Tench Coxe on June 1, 1795: "This ball of liberty, I believe most piously, is now so well in motion that it will roll round the globe, at least the enlightened part of it, for light & liberty go together. It is our glory that we first put it into motion & our happiness that being foremost we had no bad examples to follow."[3] In the 1776 Declaration of Independence, Jefferson defended a people's right to throw off a government that failed to fulfill its duty to serve them: "We hold these truths to be self-evident, that all men are created equal, that they are endowed by their Creator with certain unalienable Rights, that among these are Life, Liberty and the pursuit of Happiness. That to secure these rights, Governments are instituted among Men, deriving their just powers from the consent of the governed."[4] But would he also apply this statement to a colonial society consisting of black people? The revolution in Saint-Domingue would serve as a crucial test of Jefferson's ideals.

Humboldt on the Haitian Revolution

Alexander von Humboldt maintained throughout his life a special consciousness of fundamental principles regarding liberty, equality, and fraternity, on which he based his own philosophy. He never approved of the revolutionary fervor of the Jacobins, however, and he recoiled from the Terror of the French Revolution as well as the violent slave revolts in Haiti and elsewhere. According to Humboldt, such responses to tyranny could never allow for the construction of a progressive society. In spite of his severe criticism of colonialism and his repeated warnings of the serious consequences that could result, he never recommended a military solution, as he rejected all forms of violence used to accomplish political ends. Instead, convinced that many of the major problems in the New World originated in the colonial administrations, he made concrete propositions regarding extensive reforms in the governments and established institutions. In Humboldt's eyes, the hate and discord in the colonies was a by-product of rule by European governments, leading to a confusion of ideas and sentiments that could result in a general revolution, though in most cases, he noted, the mutinous indigenous people confined themselves to running the Europeans out of the country and starting a war among themselves.[5]

Humboldt had always been cautious in his comments concerning the political situation in the regions he visited in America. He did not want to interfere in that complicated sphere and preferred to avoid conflict with the local authorities. Nevertheless, his travel journal contains frank comments on the first signs of a desire for liberty among the people of a territory he visited. It seems odd that no mention of the Haitian Revolution is made in the published part of his journals, even though Haitian independence was proclaimed in the very year his American expedition ended, 1804. Such documents may still be discovered; a few years ago, twenty-four pages of Humboldt's travel journal with unknown annotations concerning Cuba were found in the Biblioteka Jagiellonska of Cracow.[6] This material carries the title "Isle de Cuba. Antilles en general" and purportedly contains information relevant to Saint-Domingue.[7]

Neither do Humboldt's books offer treatises or long reflections about the occurrences in Saint-Domingue; these contain only a few brief comments about the danger the revolution might prefigure for the rest of colonial America. His two regional essays about New Spain and Cuba also contain remarks of this nature; in his travel description, *Personal Narrative,* however, no such reference can be found. Humboldt's essay on New Spain mentions the "tragic events" of the Caribbean revolution and regrets that the fear they caused surmounted other issues: "So true it is, that the fear of physical evils acts more powerfully than moral considerations on the true interests of society, or the principles of philanthropy and of justice, so often the theme of the parliament, the constituent assembly, and the works of the philosophers."[8] However, he seems rather optimistic about the effect of these "tragic events" on America, noting that they must surely lead to a reduction in the traffic of slaves. It is primarily in Humboldt's treatise on Cuba that he integrates the results of the Haitian Revolution into his own argument. Humboldt had devoted much thought and energy to the struggle against slavery, particularly in Cuba but also in the United States, proposing numerous reforms designed to reduce and finally eliminate it. Knowing that arguments based on humane concerns would not be convincing to all people, he used statistical and economical data to prove that free laborers were more productive.

As a result of Humboldt's warnings, the Haitian Revolution was at last being interpreted as a cautionary tale for Cuba. His essay on this island and particularly his chapter on slavery sound the alarm: "If the legislation of the Antilles, and the condition of the colored population, does not soon experience some salutary change, and if discussion without action is continued, the political preponderance will pass into the hands of that class which holds the power of labor, the will to throw off the yoke, and valor to undergo great privations. This bloody catastrophe will occur as the necessary consequence, of circumstances, and without the free negroes of Haiti taking any part whatever, they continuing always the isolated policy they have adopted."[9] Humboldt sharply criticizes the position of the white population, which he believed did not understand the urgency of the matter, "for simultaneous action on the part of the negroes, seems to them impossible, and every change, or concession made to a population subject to servitude, is deemed to be cowardice. But it is not yet too late, for the horrible catastrophe of St. Domingo happened because of the inefficiency of the government."[10] In calling the circumstances in Haiti "horrible" and a "bloody catastrophe," he refers not to the results of this movement, which he believed to be a "necessary consequence from the circumstances," but to the violent process itself, which, for one who firmly believed in the transformative power of Enlightenment ideals, had to be avoided at all costs. He strongly endorsed the recognition of the state of Haiti by the French government, at least as a partial solution to the problem.[11] Humboldt appears to be content with the progress made in Haiti during the twenty-six years following the first revolution, not only for the inhabitants of the island, but also for the white men of the Spanish and English dominions nearby. However, he warns about a "fatal security, which disdainfully resists any improvement in the state of the servile class," since, as he notes in the conclusion of his book, "the fear of danger will force those concessions which the eternal principles of justice and humanity demand."[12]

In summary, while Humboldt does not offer a profound study of the Haitian Revolution or speculate regarding its causes and the consequences, he does apply lessons learned from the events in Haiti to the situation in Cuba. He describes revolutionary incidents and the fears they have incited among the white population and integrates them into his own arguments against slavery. Thus Humboldt's interest in slavery dates back not to Cuba, but to the revolution in Saint-Domingue.[13] His essay can be understood as an appeal to the enlightened intellectuals of the Cuban elite, with whom he associated during his stay on the island. Among these, Francisco de Arango y Parreño should be particularly singled out. A politician, economist, intellectual, and leader of the first reformist movement in Cuba at the end of the eighteenth and beginning of the nineteenth century, he became Humboldt's longtime correspondent.[14]

Jefferson on the Haitian Revolution

Thomas Jefferson's attitude toward the events in Saint-Domingue was considerably more complex than Humboldt's and more conditioned by the political and social circumstances in which he lived. Several important elements inflected his political thought: the institution of slavery, his personal relationships with black people, the use of revolution as a political expedient, his ideas regarding the future of the New World in general, as well as his doubts regarding the applicability of democratic political structures to other American regions. These considerations rendered the events on the French island a complicated issue for him. Furthermore, since Jefferson was president of the United States during 1801–9, his actions were necessarily guided by a constellation of divergent interests. Those years saw major changes in the foreign and domestic politics of his country that influenced Jefferson's political and philosophical positions. Thus, in order to understand the evolution of Jefferson's political attitude toward Saint-Domingue, it is necessary to place him in the proper historical context.[15]

From 1791 onward, concomitant with the first signs of revolt, Jefferson expressed concern about the possibility of attendant danger. His initial reaction was to issue a proposition for the gradual emancipation of Haitian slaves. In a 1799 letter to James Madison, he bemoaned the risk of an insurrection inspired by Saint-Domingue and concluded that "against this there is no remedy but timely measures on our part to clear ourselves by degrees of the matter on which that lever can work."[16] In his position as head of the government, he favored joining forces against the regime of Toussaint Louverture,[17] but when he became aware of the broad ambitions of Napoleon in America, he changed his opinion and reneged on his promise to aid the French.[18]

During the war years of 1802–3, Jefferson officially followed a policy of neutrality, a position that caused problems for the French, who did not have sufficient means for a military expedition and depended on American support. Jefferson's "neutrality," however, actually consisted of considerable aid to the regime of Toussaint Louverture in Haiti, which must be understood in connection to the Louisiana Purchase. When the army of Napoleon Bonaparte's brother-in-law Charles Leclerc landed in Haiti en route to New Orleans in late 1801, dispatched in an attempt to regain French control of the island, Jefferson did all he could to support Toussaint Louverture to fight the occupation army.[19] The idea was that if the Haitians defeated the French soldiers, fewer of the latter would be available to defend New Orleans. With the French threat neutralized after the acquisition of Louisiana in 1803, Jefferson changed course and began to see the Haitian leader as a danger. Although he encouraged the independence of Saint-Domingue, he refused to recognize the new regime, fearing that the Haitian Revolution would inspire similar action among the slaves in his country. The United

States' rejection of diplomatic relations with Haiti would continue until 1862. Jefferson even proclaimed an embargo on trade with the island. When, before the proclamation of independence in June 1803, Dessalines sent him a letter trying to reestablish commercial connections and friendly relations with the United States,[20] Jefferson did not bother to reply.[21] This position ran counter to the attitude he adopted as secretary of state in 1792, when he defended recognition of the French revolutionary government. It also ran counter to the principles of the American Declaration of Independence, which established the right of the people to reject an oppressive government: "That whenever any Form of Government becomes destructive of these ends, it is the Right of the People to alter or to abolish it, and to institute new Government, laying its foundation on such principles and organizing its powers in such form, as to them shall seem most likely to effect their Safety and Happiness. . . . But when a long train of abuses and usurpations, pursuing invariably the same Object evinces a design to reduce them under absolute Despotism, it is their right, it is their duty, to throw off such Government, and to provide new Guards for their future security."

In order to understand Jefferson's attitude toward Haiti, it is important to recognize the relationship between the United States and France, and particularly, the interests of his young nation in this relationship.[22] Jefferson's sympathy for the French inclined him to renounce the principle of neutrality that he had established as the foundation for North American foreign policy. This sentiment echoes the principle that "the enemy of my enemy is my friend." But neutrality was also very advantageous for the United States, especially economically. In the case of Haiti, between 1803 and 1812 the United States was the island's most important trading partner, particularly in comestibles.[23] Also influencing U.S. relations with France, and thus the issue with Saint-Domingue, was the U.S. government's ambition to obtain West Florida after the purchase of the Louisiana Territory.[24] The Haitian embargo was renewed automatically until 1810, when the acquisition of West Florida became certain. With French support suddenly no longer indispensable, the United States could resume its fruitful traffic with the formerly shunned French colony.[25]

The reaction of slave owners in the southern United States was yet another factor that influenced Jefferson's political attitude and actions toward Saint-Domingue, especially after the slave conspiracies in the region around Richmond. The Haitian Revolution inspired a first slave insurrection in Virginia in 1800, Gabriel's Rebellion, named after its leader, a literate and highly skilled blacksmith named Gabriel Prosser.[26] The ideological basis for the attempted rebellion was the Declaration of Independence, with the rebels maintaining that "liberty" applied to blacks as well as whites. In 1802, one of Prosser's followers, Sancho, led a second failed insurrection called the "Easter Plot." When the preparations for this revolt were discovered, great fear spread throughout the white population of Virginia, initiating a campaign of terror against both actual and suspected insurrectionists.

Jefferson displayed some comprehension that slaves were motivated by the same ideals that had inspired the American colonists to rebel against their British masters, but this was far outweighed by his fear of possible future uprisings. His loyalty toward the planters of Virginia was fierce—he was one of them, after all—and he could understand their unease as well as their racial phobias. He felt strongly obligated to defend what they considered to be their rights. In a letter to Rufus King, the U.S. minister to Britain, Jefferson described the situation in Virginia and emphasized the vulnerability of the American South: "The course of things in the neighboring islands of the West Indies appears to have given a considerable impulse to the minds of the slaves in different parts of the US. a great disposition to insurgency has manifested itself among them, which, in one instance, in the state of Virginia broke out into actual insurrection. this was easily suppressed, but many of those concerned, (between 20. & 30. I believe) fell victims to the laws."[27] Jefferson's analysis of the independence movements in Spanish America helps to illuminate his position in this particular context: the real difficulty, according to him, was not merely to overthrow a government, but to successfully establish a new one, based on republican principles, in its place.

As a consequence of Gabriel's Rebellion—or, as Jefferson called it, "the tragedy of 1800"—he proposed in an 1801 letter to James Monroe that Saint-Domingue be designated a home for deported slaves and free blacks. After considering other possibilities, such as purchasing land for them within the western territory of North

America, or within the Spanish dominions, he came to the conclusion that

> the West Indies offer a more probable & practicable retreat for them. inhabited already by a people of their own race & colour; climates congenial with their natural constitution; insulated from the other descriptions of men; Nature seems to have formed these islands to become the receptacle of the blacks transplanted into this hemisphere. whether we could obtain from the European sovereigns of those islands leave to send thither the persons under contemplation, I cannot say: but I think it more probable than the former propositions, because of their being already inhabited more or less by the same race. the most promising portion of them is the island of St. Domingo, where the blacks are established into a sovereignty de facto, & have organised themselves under regular laws & government.[28]

Following his idea of expatriating blacks as a solution to slavery and a means of quieting racial fears, he also proposed to Rufus King that a group of insurgent slaves could be deported to West Africa under the auspices of the Sierra Leone Company, an English abolitionist organization that had established Freetown as a home for former slaves. Unfortunately, the negotiations with the company were not successful, and most of the accused conspirators were sold as slaves to Spanish and Portuguese colonists.[29] Jefferson was also obliged to take into consideration the interests of expansionists, particularly after the purchase of Louisiana. The fact that in 1804 he did not oppose the extension of slavery into the new territories may also shed some light on his attitude toward this Caribbean island. Jefferson was conscious that his policy toward Haiti was not very coherent and that the changes in his politics were largely dictated by domestic considerations. He was a man torn between his own convictions and the political conditions of his time, trying to reconcile sharply antagonistic positions in order not to risk disunion in the country, while also representing its geopolitical interests. He found his humanist convictions and principles being subordinated to international interests and the economic ambitions of U.S. expansionism. His attitude and political behavior suggest that he cared less about Haiti and its revolution than his enlightened views might lead one to expect. He appeared to be more interested in the island as a chess piece he could deploy in a tactical game against Napoleon, something that might be discarded it when it was no longer of use.

It would be interesting to know whether Humboldt and Jefferson discussed the situation in Haiti when they met in Washington in the spring of 1804. Their correspondence over the next twenty-one years shows no record of such a conversation, despite the fact that they frequently discussed political developments. This may seem surprising, but, as noted, the two avoided several important topics in their communication, probably because they held extremely dissimilar positions and did not wish to strain their relationship with disagreement. Nevertheless, in Humboldt's correspondence with William Thornton, in which he also made references to slavery, he told Thornton that he had read his publication on the institution of slavery, "Political Economy" (1804), stating that the "more the events there have offended the truth, the more it seems to be the obligation of a moral person to put this problem to an end."[30]

As a scientist, Humboldt could approve the aims and the political results of the Haitian Revolution—rejecting, of course, its violent aspects—and express this opinion publicly since the revolution's goals and outcome accorded with his own convictions. Jefferson's political position, on the other hand, required him to consider all possible responses in light of the interests of his country, the reactions of his enemies, and the influence of his actions on the forthcoming elections. This was his reality as an important figure in the creation of a new society. The diverse reactions that the Haitian Revolution provoked in Humboldt and Jefferson cannot be considered independently from the many other topics they discussed. Their responses reveal personal convictions, but they are equally linked to historical circumstances, political interests, and private concerns.

7 Engagement with the Natural World

The eighteenth century—when Thomas Jefferson and Alexander von Humboldt were growing up—and the beginning of the nineteenth century were characterized by a seemingly endless series of discoveries and innovations, of which the development of new social and political structures was just a small part. In the broad field of natural history—defined for present purposes as the systematic study of any category of natural objects or organisms in their environment, based on observation rather than experimental methods—important changes were taking place. Innovative approaches to measuring and studying nature according to the scientific principles of the Enlightenment, and to establishing an order for all living beings led to biological classification systems. A deeper and more nuanced understanding of the natural environment was developing. The influence of the climate on human beings and man's interaction with the natural world were much-debated issues. Once again, Humboldt and Jefferson—both deeply interested, and conditioned by their personal interests and backgrounds—showed distinctive approaches.

The European encounter with America resulted in fundamental changes in the study of both nature and ethnography in the New World. The task now was to incorporate the newly discovered information into what was already known of the Old World, and to establish schemes into which both might fit. By the end of the eighteenth century, many a colonial territory had begun to search for an individual identity, a process likely moving toward political independence that cannot be seen as separate from the debate on nature. The understanding of the interconnection between nature and humans was influenced by multiple factors, including religious convictions, Eurocentric beliefs, and nationalistic considerations. It further depended on the type and reliability of information available, and whether one was from the Old or New World.

In the early eighteenth century, the meaning of the term "natural history" was quite different from what it is today. Natural history was bipartite, comprising natural philosophy (looking for definition, description, and material causes of nature and the physical universe) and moral philosophy (focusing more on questions of morality and metaethical discussion of the nature of moral judgments, meanings, and values).

The study of natural history is considered to begin with Aristotle and other classical philosophers who analyzed the diversity of the natural world. Aristotle dedicated himself to the study of animals and classified them according to their method of reproduction; he believed that living beings were arranged on a ladder of perfection rising from the most lowly plants to man. Together with Plato he developed the concept of the *scala naturae,* or great chain of being, organizing life forms into a strict hierarchical structure of life, believed to be of divine origin. The chain starts with God and the angels beneath him, moving downward to include mankind, animals, and plants, followed by minerals. Another early contribution was made by the Roman natural philosopher Pliny the Elder, whose encyclopedic work *Historia naturalis,* published around 79–77 BC, became a model for all subsequent such works. From the work of these ancient scholars until that of Carl Linnaeus and other eighteenth-century naturalists, the principal ordering concept of natural history remained the *scala naturae.*

During the Enlightenment, rapidly expanding knowledge of natural history and the increase in the number of known species generated interest in the idea of imposing a general system of order upon the many natural-history collections being gathered. Several attempts to identify, classify, and organize the species into taxonomic groups culminated in the universally accepted system established by Linnaeus. He abandoned the long descriptive names of classes and orders used by his immediate predecessors and introduced the consistent application of binomial nomenclature. All organisms were given two Latin names: one for the genus and one for the species. This system, too, was hierarchical, starting with three kingdoms (mineral, vegetable, and animal), which were divided into five ranks: classes, orders, genera, species, and sometimes, below the rank of species, taxa named *variety* (botany) or *subspecies* (zoology).[1] His most famous work, *Systema naturae* (1735), is the foundational work of zoological nomenclature; likewise his *Species plantarum* (1753) gave birth to modern botanical nomenclature. He described specimens on the basis of physical appearance and manner of reproduction, and classified them relative to each other according to their degree of similarity. Linnaeus was the first to classify humans among the primates, observing that both species had the same basic anatomy and that outside of the human ability to speak, there were no major differences. He placed man and monkeys in the category *anthropomorpha*. Unsurprisingly, this classification generated considerable theological criticism. Many felt that man was being pushed off the spiritually higher plane he had been presumed to occupy in the great chain of being. If, according to the Bible, man was created in the image of God, would this apply to monkeys and apes as well? This conflict between the scientific versus theological understanding of nature would be revitalized when the publication of *On the Origin of Species* by Charles Darwin in 1859 initiated the creation-evolution controversy.

Though Linnaeus established a precise system for the classification of nature, he contributed little analysis or interpretation. This is to be expected since Linnaeus believed that he was simply revealing the unchanging order of life as it had been created by God. His religious beliefs convinced him of the invariability of species whose numbers and characteristics had not changed throughout history. Following this logic, species in similar environmental conditions but different locations ought to be indistinguishable, but this soon proved not to be the case. By the end of the eighteenth century, an increasing number of scientists began to question his understanding of nature. Instead of the static and harmonious world he envisioned, they observed nature to be highly mutable. Among Linnaeus's opponents was the Comte de Buffon, who criticized the Linnaean taxonomy in his work *Histoire naturelle, générale et particulière*.[2] Although he asserted that species were able to change over generations as a result of influences from the environment, he rejected the idea that species could evolve into other species. One of his most significant contributions to the biological sciences was his insistence that natural phenomena had to be explained by natural laws and not by theological doctrine.

These early ideas regarding the possibility of evolution among living beings were developed further by Erasmus Darwin, the grandfather of Charles Darwin. The first evolutionist who publicly stated his ideas about the processes leading to biological change—though his theory about these processes was not correct—was Jean-Baptiste Chevalier de Lamarck, a protégé of Buffon. It was the French scientist George Cuvier who discredited Lamarck's first theoretical framework of organic evolution, advocating instead the theory of catastrophism, which held that violent and sudden natural catastrophes such as great floods and the rapid formation of major mountain chains had led to the extinction of living beings and to the subsequent evolution of new species. Nevertheless, a careful examination of European geological deposits in the early nineteenth century undertaken by the English geologist Charles Lyell proved Cuvier's theory to be wrong, indicating that the changes had occurred more slowly and progressively. He provided conclusive evidence for the theory of uniformitarianism, developed originally by the late-eighteenth-century Scottish geologist James Hutton, which posited that the natural forces changing the shape of the earth's surface had been operating the same way in the past. This idea was instrumental in Charles Darwin's development of his theories on biological evolution in the 1830s. Darwin's theory of evolution by natural selection provided a mechanism not only for understanding how species arose, but also for interpreting patterns in the distribution and abundance of species. A central argument of his work *On the Origin of Species* was that plants and animals had the potential to reproduce very quickly and reach huge population densities, though this potential was rarely achieved because each species was subject to a series of natural checks and balances. Darwin's work thus made an important contribution to the development of ecological sciences at a moment when there was still no term for the field he had created. It was the German biologist Ernst Haeckel (1834–1919) who first introduced the word "ecology" in 1866 in his *Morphology of Organisms*.

Ecology is the most comprehensive and diverse science, yet it is one of the youngest disciplines, and in spite of several attempts to describe its origins and basic concepts, its history is not well defined.[3] The relationship between climate and vegetation is among the earliest ecological observations. As early as the third century BC, the Greek philosopher Theophrastus, a student of Aristotle and Plato, classified more than five hundred plants into major growth forms (trees, shrubs, herbs) and further organized them according to morphology. He conducted experiments by transplanting species to areas outside their natural range to determine if they would grow, and documented systematic changes in patterns of deciduousness and evergreenness with distinct climate conditions. He also observed the positive relationship between altitude and latitude with respect to climate and vegetation.[4] More than two thousand years later, the next wave in the history of ecology was spurred by the scholars of the sixteenth and seventeenth centuries, and their observations were built upon by the great botanist-explorers of the eighteenth and nineteenth centuries, some still under the influence of the ideas of the ancient Greek thinkers.[5] This is the time when Humboldt and Jefferson appear on the world stage of science.

An Approach from the Old World

From the beginning of his scientific activities, Alexander von Humboldt clearly defined the aim of his research: to study, analyze, and describe the natural world. He insisted that the only way to understand nature's complexity was to take accurate measurements in the field and then search for general laws. His concept of science envisaged the earth as an organic whole, all parts of which were mutually interdependent. Therefore nothing in nature could be studied in isolation. In keeping with this holistic view, Humboldt, from his early years onward, looked for the interconnection of humans and the natural world.[6] By 1793—before his famous American expedition—he had this idea clearly in mind when he summarized the scientific interest in the definition and methodological explanation of what he by then called the *physique du monde*, a universal science. This idea culminated in his last work, *Cosmos*, presented as the "physical description of the universe." Between 1795 and 1799, Humboldt projected the elaboration of a complete geography of plants of the world that would be connected with the geophysical forces he identified and oriented toward the "chain of connection" instead of being limited to "mere encyclopedic aggregation."

Any discussion of Humboldt's concept of plant geography must call to mind his famous mentor Carl Ludwig Willdenow, considered to be one of the first scholars of phytogeography, the study of the geographic distribution of plants. In his position as director of the Botanical Garden in Berlin, Willdenow was occupied with taxonomic studies based on collections from various parts of the world sent to him by other naturalists. The extensive herbarium he created included more than twenty thousand species, including many plants that had been newly discovered and described. The collection was purchased after his death by the Botanical Garden in Berlin, which had greatly prospered under his directorship and still exists today. Willdenow was interested in the adaptation of plants to different climate zones and also in the fact that the same climate produced different species of plants with common characteristics. Throughout his American expedition, Humboldt stayed in contact with Willdenow, corresponding with him and later sending him vast quantities of material gathered in the New World. Willdenow's phytogeographical studies led to ideas that were later developed further by Humboldt and others. Among these concepts were the conviction that plant distribution patterns changed over time, that climate influenced the number as well as the character of plant species in particular geographic areas, and that it was possible for new plant species to develop and existing ones to become extinct. Willdenow published these thoughts in 1792 in his work *Grundriss der Kräuterkunde* (Basics of herb science), which laid the groundwork for Humboldt's *Geography of Plants*. In the foreword of the 1807 German edition, *Ideen zu einer Geographie der Pflanzen*, Humboldt credits Willdenow's works with inspiring him.[7] During the preparation of his other botanical writings, Humboldt continued to collaborate with his former teacher, sending him plants for study purposes. In a letter written in May 1810, he invited Willdenow to Paris to work with him on the publication *Nova genera et species plantarum*. Unfortunately, after working with Humboldt for only a few months on his South American herbarium, Willdenow was forced by ill health to return to Berlin and died shortly afterward. Willdenow was replaced by one of his students, Karl Sigismund Kunth, who continued his work from 1813 onward and contributed significantly to Humboldt's botanical work, which became a classic of botanical literature. Kunth's participation in the work on Humboldt's herbarium resulted in a shift in the methodology applied. The period between 1805 and 1830 was one of transition in botany from the classification system developed by Linnaeus to a more natural system constructed by Antoine-Laurent de Jussieu and Augustin Pyrame de Candolle. Kunth, a follower of these newer ideas, used a methodology quite different from that of Willdenow, who was a devotee of Linnaeaus.[8]

In his *Geography of Plants*, Humboldt introduced an idea he went on to develop much more fully in his final work, *Cosmos*. "In the great chain of causes and effects," he observed, "no material, no activity can be considered in isolation."[9] Humboldt was especially preoccupied with the distribution of vegetation and its relationship to climatic zones, as well as other factors that affected the way plants took hold in specific regions, and less with detailed descriptions of individual plants or species. As he wrote in his *Personal Narrative:* "I was passionately devoted to botany, and certain parts of zoology, and I flattered myself that our investigations might add some new species to those which have been already described; but preferring the connection of facts which have been long observed to the knowledge of insulated facts, although they be new, the discovery of an unknown genus seemed to me far less interesting than an observation on the geographical relations of the vegetable world, or the migration of social plants, and the limit of the height which their different tribes attain on the flanks of the Cordilleras."[10] He believed that the "philosophical study of nature rises above the requirements of mere delineation, and does not consist in the sterile accumulation of isolated facts."[11] Humboldt's plant geography thus created a link between the natural sciences and the human sciences, constituting a distinctive tradition of inquiry that developed and diversified throughout the nineteenth century.[12]

In Humboldt's final and synthesizing work, *Cosmos*—his scientific opus, which articulated a grand theory of natural history—this concept was extended through the unification of all creation on earth as well as in the universe, in order to present what Humboldt called a "physical description of the world." In the book's foreword, he describes the goal in relation to his holistic idea: "The principal impulse by which I was directed was the Earnest endeavour to comprehend the phenomena of physical objects in their general connection, and to represent nature as one great whole, moved and animated by internal forces. My intercourse with highly-gifted men early led me to discover that, without an earnest striving to attain to a knowledge of special branches of study, all attempts to give a grand and general view of the universe would be nothing more than a vain illusion." Descriptive botany, he added, was no longer "confined to the narrow circle of the determination of genera and species," which leads the observer to the study of the geographic distribution of plants in any place of the world. Further, it was "necessary to investigate the laws which regulate the differences of temperature and climate, and the meteorological processes of the atmosphere, before we can hope to explain the involved causes of vegetable distribution; and it is thus that the observer who earnestly pursues the path of knowledge is led from one class of phenomena to another, by means of the mutual dependence and connection existing between them."[13] Humboldt's contribution to science, then, was not focused on one field, but on many: its particular relevance lies in the method he applied.

According to Humboldt, climate was a very important factor in the study of the geographic distribution of the plants: "It is by subjecting isolated observations to the process of thought, and by combining and comparing them, that we are enabled to discover the relations existing in common between the climatic distribution of beings and the individuality of organic forms . . . ; and it is by induction that we are led to comprehend numerical laws, the proportion of natural families to the whole number of species, and to designate the latitude or geographic position of the zones in whose plains each organic form attains the maximum of its development."[14] This conviction led Humboldt to the observation and study of climatic conditions and the measurement and comparison of temperature; with his collected data he sought to identify patterns in nature.

Humboldt sought interconnections not only between different academic disciplines but also between certain geographic regions. The comparison between territories interested him, including the connection between the political and social movements he observed there. He had long hoped to carry out a program of comparative studies between America and Asia, although financial considerations limited his Asian expedition to his travel through Russia in

1829. His view was thus necessarily more directed toward the relationship between Europe and America. As a consequence, though for him our modern term "Atlantic world" did not exist as such, he was completely aware of the particular interconnectedness of the regions that form part of this world.

Humboldt frequently stressed the importance of personal observation and the study of nature in its actual surroundings rather than in a collection or museum as other naturalists of his time might do. In fact, it was quite common for so-called "armchair scientists" to gather their data from research undertaken and published by others, and to construct their theories based on such findings. Humboldt's innovative approach, on the other hand, combined fieldwork with a scholarly elaboration of the results. He even performed certain scientific experiments on his own body, testing, for example, the contraction of his muscles in response to electricity; the effects of a small dose of a poison taken in an Indian village at the Orinoco River; and his physical reactions to extreme height in the mountains of Ecuador.

In regard to both his epoch and his working methods, Humboldt must be situated between Enlightenment values and Romanticism. His scientific concepts conform to the postulates of the Age of Reason, as is evident in his use of measurement instruments to explore and understand the unknown world. He kept field notebooks; numbered and classified the specimens he found; prepared, together with his collaborators, numerous scientific illustrations; and published his research results with strict attention to procedure and detail. On the other hand, his integrative and global vision of the American reality led him to more general considerations that anticipated the Romantic period, which integrated subjective perception into the description of nature.[15] A gradual change in his methods can be observed over the course of his voyage. During the time Humboldt spent on the Iberian Peninsula, his interests were focused on the Enlightenment paradigms of science. With his arrival in Tenerife, however, his view seemed to expand and take into account other elements of the journey, such as the beauty of the landscape, the character of the inhabitants, their cultural and literary achievements, and so forth.[16] In the Canary Islands, he began to develop the integrated vision that would characterize his exploration of the New Continent.

Humboldt did not separate science and art, but used them to complement each other.[17] The results of his American expedition were presented in both written form and through beautiful illustrations of landscapes and other elements of nature, both floral and faunal. Through visual representations of scientific information he sought to make more palpable the links between different phenomena. Among his more famous illustrations are two mountain cross-sections, one representing the Teide, the highest mountain in Spain, situated on Tenerife, and the other the Chimborazo in Ecuador, where he showed the distinct "zones of habitation"—the kinds of plants that grow at a given altitude under particular climatic conditions.

As mentioned, Humboldt's holistic scientific concept included humankind among the interconnections and interdependencies of the components of nature. With his fundamental assumption that neither humans nor nature could be understood in isolation, that the human being was conditioned by the environment he lived in, Humboldt made an important contribution to ecological discourse.[18] Nature played an essential role in his writings, and not only in those publications dedicated to the representation of natural phenomena. His exploration of the different landscapes on the American continent was fundamental in the formulation of all his theories and scientific convictions. As a consequence of what he had seen and analyzed early in his expedition, Humboldt was able to argue, in his *Geography of Plants,* that cutting down forests causes climate change. His delineation of the consequences of deforestation (exposing the bare soil to heat and wind) and the damage caused by Europe's exploitation of its tropical colonies (through water shortage or mono-agriculture, an idea he develops further in his *Political Essay on the Kingdom of New Spain*) can be considered important early steps in the field of environmental science. Discussing the aridity of the central Mexican plains and the lack of trees in his work on New Spain, for instance, he states: "These disadvantages have augmented since the arrival of Europeans in Mexico, who have not only destroyed without planting, but in draining great extents of ground have occasioned another more important evil."[19] Along the same lines, he also criticizes the poor irrigation system established by the Spanish in Mexico:

> This diminution of water experienced before the arrival of the Spaniards, would no doubt have been very slow and very insensible, if the hand of man, since the period of the conquest, had not contributed to reverse the order of nature. Those who have travelled in the peninsula know how much, even in Europe, the Spaniards hate all plantations, which yield a shade round towns or villages. It would appear that the first conquerors wished the beautiful valley of Tenochtitlan to resemble the Castilian soil, which is dry and destitute of vegetation. Since the sixteenth century they have inconsiderately cut, not only the trees of the plain in which the capital is situated, but those on the mountains which surround it.[20]

Also in his *Personal Narrative,* he mentions that "the first colonists very imprudently destroyed the forests," and as causes of the diminution of the lake of Valencia, he enumerates again the destruction of the forests, the clearing of the plains, and the cultivation of indigo, among other factors.[21]

It is interesting that although he normally tended to limit his specific criticism of the policy of the Spanish government to its oversea dominions, in the context of these ecological concerns he became very direct about his rejection of the colonial exploitation of other countries. Here he offered his first warnings that remaking the landscape would affect the delicate natural balance, which would lead to significant natural destruction that would impact human beings, who form a part of the natural system. In his determination to examine the consequences of human activities on the natural environment, Humboldt might be characterized today as thinking globally. In his subsequent analysis, he combined the descriptive approach to the field of natural history with the quantitative and conceptual understanding of natural philosophy and thus made an important contribution to establishing the field of *biogeography,* the study of the distributions of organisms in space and time.

Several scholars have recently focused more intensively on the ideological groundwork Humboldt laid for the new areas of study that are now called now climatology and ecology. The Prussian scientist's environmental thinking preceded and inspired that of the American naturalists Henry David Thoreau, George Perkins Marsh, and John Muir; Humboldt's work particularly influenced Thoreau, who classified New England's climate zones according to Humboldt's concept of plant ecology.[22] Thus, inspired by a Prussian explorer more than a hundred years before the idea of an ecosystem caught on in the popular imagination, many of America's first naturalists and scientists were laying the groundwork for this new branch of science in the United States.[23]

In the context of today's heightened concerns over global change, human-induced crises, sustainability, and conservation of the environment, Humboldt's holistic vision of human interdependence with nature assumes renewed relevance. With his holistic thinking and his work on the geography of plants, Humboldt made an important contribution to the development of the ecological sciences. He was an intellectual giant of his time, a celebrated and prolific scholar. Nevertheless, his merits should not eclipse the contributions made by his fellow scientists. There is still debate among historians regarding the origins of ecological thinking, and no particular person, date, or occurrence is generally considered to mark the beginning of it. Though its fundamental concerns can also be found in ancient Greek thinking, it is reasonable to say that ecology gradually emerged as a distinct discipline during the second part of the nineteenth century from a diverse array of previously investigated areas, including plant geography, taxonomical classifications, and Darwin's theory of evolution. Like many disciplines, the sciences of ecology were not based on the work of a single individual but developed as a chain of ideas, with scholars building on the ideas of others. Humboldt's defining contribution to ecology may be that he provided a solid link between Linnaeus and Darwin.[24]

An Approach from the New World

Jefferson's approach to the natural world differed considerably from Humboldt's, since he was born and raised at the edge of the Virginia frontier and from early childhood was surrounded by the world of nature, virtually unaltered by mankind. He loved Virginia, the richly varied landscapes, its flora and fauna, and yearned to be at home when he was away. His first home was Shadwell, the farm of his father, Peter Jefferson, and it was there that in his early years he became a collector of minerals, plants, animal bones, insects, and fossilized shells. The young Jefferson was fascinated by nature and became its close observer. He maintained a strong lifelong connection to his natural environment, which provided him with personal tranquility and balance. As he famously wrote to Pierre Samuel du Pont: "Nature intended me for the tranquil pursuits of science by rendering them my supreme delight. But the enormities of the times in which I have lived, have forced me to take a part in resisting them, and to commit myself on the boisterous ocean of political passions."[25] During his years representing the young American nation in Paris, Jefferson's enthusiasm for the natural sciences was reinvigorated through contact with the leading scholars there.[26]

There were two sides to Jefferson's attitude toward the natural world. Many aspects of wild nature appealed to his aesthetic sensibility, to his heart, and he always associated his greatest contentment with closeness to nature. In numerous letters, especially those directed to women, and among them particularly to his close friend Maria Cosway, Jefferson could wax very eloquent about landscapes. In 1774, he purchased the 157 acres surrounding Natural Bridge near Lynchburg, one of his favorite places in Virginia, which he called "most sublime of Nature's works."[27] When he visited the spot for the first time in 1767, he sketched the bridge and made annotations in his memorandum book, which served as the basis for his commentary in *Notes on the State of Virginia,* the first book on the natural history of Virginia. His description of Natural Bridge not only provides information about its size and geological formation, but also records the impression the natural wonder made on him: "It is impossible for the emotions, arising from the sublime, to be felt beyond what they are here: so beautiful an arch, so elevated, so light, and springing, as it were, up to heaven, the rapture of the Spectator is really indescribable!"[28]

But Jefferson was a practical man, too, and he lived in an age when nature was seen as something to control, shape, and change. Taming nature meant acquiring the detailed knowledge to understand its mechanisms,[29] which he did by working with it day after day.

Jefferson had an extensive library of books on the natural world, including publications by the best-known natural historians of his time—Buffon, Linnaeus, Smith Barton, Cuvier, Humboldt—and was greatly inspired by their works.[30] In his own writings, the word "nature" appears frequently and in diverse contexts: natural law, natural right, American nature, natural reason, natural means, and so forth.[31] For Jefferson, the best knowledge was practical, since in early America pragmatic matters were often pressing and required immediate solutions. In a letter to John Adams, he remarks, "I am not fond of reading what is merely abstract, and unapplied immediately to some useful science."[32] For Jefferson, natural science was never "merely" abstract: it related to many tasks he carried out as a farmer, and even as a philosopher and politician. The most useful and satisfying lives, he believed, were those lived close to nature, and thus he believed that country dwellers should form the basis of a democratic nation. This idea stood in sharp contrast to the European Enlightenment approach to nature, or that prevailing during the Romantic era. Jefferson's image of the countryside can be understood as a form of nationalism: against Europe's rich history and civilization, he placed America's sublime nature, which provided the new country with a base from which to prosper.[33] Even in his last years, when he was selecting the classes to be taught in the University of Virginia, he championed the causes of nature: out of its eight schools, two were devoted to natural history and natural philosophy.

Like his Prussian counterpart, Jefferson wanted to understand the world in part through scientific measurement. His delight in measuring instruments began in his childhood, when he became familiar with surveying through his father. Years later, when he was based in Europe, he took the opportunity to purchase the latest instruments of measurement and began using them for his field studies in America.

Among the numerous scientific disciplines that interested him, meteorology held a special attraction. He was a rabid note taker and list maker: In addition to a farm book, he kept a weather memorandum book, in which from 1776 almost until his death he noted temperature, rainfall, barometric pressure, and wind directions, always eager to contrast those data with measurements taken by others.[34] In order to establish these comparisons, he encouraged a network of persons to observe the weather and collect data at distant places. Thus Jefferson was the first in the American colonies to conduct systematic and detailed meteorological studies, maintaining records of all forms of climatologic phenomena and data: "My method is to make two observations a day, the one as early as possible in the morning, the other from 3. to 4. aclock, because I have found 4. aclock the hottest and day light the coldest point of the 24. hours. I state them in an ivory pocket book in the following form, and copy them out once a week."[35] He measured the precipitation and recorded the daily temperature range, but his attempts to collect data on winds and humidity were unsuccessful, owing to the limitations of his instruments. Jefferson continued to discharge these tasks for more than fifty years, a systematic weather observer to the end of his life.[36] Converting Monticello into America's first weather observation station, he was determined to understand the American climate, form the foundation of a reliable theory on it, and correlate the data with such periodic phenomena as the breeding and migration of birds and the appearance, flowering, and fruiting of plants. His interest in whether the large-scale cutting of forests could cause changes in climate was ahead of his time; and he investigated the causes and effects of the flood of 1771 and the great snowfall of the following year. In short, he had a lifelong obsession with weather in all its manifestations.[37] His interest in keeping track of the snowfall in Virginia led to the first recorded debate on global warming in history. In *Notes on the State of Virginia,* he wrote:

> A change in our climate however is taking place very sensibly. Both heats and colds are become much more moderate within the memory even of the middle-aged. Snows are less frequent and less deep. They do not often lie, below the mountains, more than one, two, or three days, and very rarely a week. They are remembered to have been formerly frequent, deep, and of long continuance. The elderly inform me, the earth used to be covered with snow about three months in every year. The rivers, which then seldom failed to freeze over in the course of the winter, scarcely ever do so now. This change has produced an unfortunate fluctuation between heat and cold, in the spring of the year, which is very fatal to fruits. From the year 1741 to 1769, an interval of twenty-eight years, there was no instance of fruit killed by the frost in the neighborhood of Monticello. An intense cold, produced by constant snows, kept the buds locked up till the sun could obtain, in the spring of the year, so fixed an ascendency as to dissolve those snows, and protect the buds, during their development, from every danger of

returning cold. The accumulated snows of the winter remaining to be dissolved all together in the spring, produced those overflows of our rivers, so frequently then, and so rare now.[38]

Nevertheless, these ideas were defeated by Noah Webster, the lexicographer, textbook pioneer, and author of the famous *American Dictionary of the English Language*, published in 1828. In 1799, Webster presented his arguments against Jefferson's conclusion in a paper read before the newly created Connecticut Academy of Arts and Sciences.[39] He considered Jefferson's contention to be an "unphilosophical hypothesis" and pronounced himself in favor of a more accurate gathering of data, criticizing the information provided by Jefferson, for lack of precision measurement instruments, as well as the unscientific method based on observations made by others. The impact of men on nature, such as clearing forests, might have led to some microchanges in climate, he sniffed, for instance, more windiness and more variation in winter conditions, but these could not be interpreted as indicators of a general weather change. While this particular controversy ended there, and though Jefferson did continue to collect weather-related data, he never again attempted to make the case for global warming.

Jefferson was also interested in astronomy and possessed an impressive assemblage of astronomical instruments. He frequently described his observations, such as the solar eclipse of June 1788, to his correspondents. He constantly promoted the use of astronomical observations for accurately establishing positions in the mapping of the country and its boundaries, as he mentions in a letter to Governor Wilson Cary Nicholas on April 19, 1816: "Measures and rhumbs taken on the spherical surface of the earth," he explains, "cannot be presented on a plane surface of paper without astronomical corrections, and paradoxical as it may seem, it is nevertheless true, that we cannot know the relative position of two places on the earth, but interrogating the sun, moon and stars."[40]

Botany was another field of deep interest for Jefferson: "Botany I rank with the most valuable sciences, whether we consider its subject as furnishing the principal subsistence of life to men and beast, refreshments from our orchards, the adornment of our flower borders, shade and perfume of our groves, materials for our buildings, or medicaments for our bodies."[41] Throughout his life, Jefferson studied both useful and decorative plants, and his achievements were honored by his fellow scholars. In 1792, the botanist Benjamin Smith Barton proposed the name *Jeffersonia diphylla* for a plant in consideration of Jefferson's knowledge in the fields of natural history.[42] Despite a lifelong career in public service, Jefferson remained a pragmatic practicing farmer. He was also horticulturist, experimenting with many varieties of plants and vegetables and converting Monticello as well as his other estates into progressive experimental farms where new plants were introduced and nurtured. He was convinced that the introduction of new plant species would direct nature for man's benefit. "The greatest service which can be rendered any country is, to add an useful plant to its culture," he maintained.[43] His interest in the useful application of knowledge in this field was manifested, for instance, in his aspiration to make wine in Virginia.[44] During his years in Europe he had taken advantage of the opportunity to broaden his knowledge of viniculture: in 1787, he undertook a tour of French and Italian vineyards, and one year later he traveled through the wine-making regions of Germany, where he closely inspected soil and climate. Unfortunately his experiments came to nothing; successful viniculture would not arrive in Virginia until the development of modern pesticides. Besides importing several types of plants to the United States with varying degrees of success, he introduced livestock such as merino sheep and sheepdogs.

We can say that Jefferson approached natural history as a scientist, but it was through gardening that he was able to participate in the rhythms of the physical world.[45] His interest in his gardens was connected to the agricultural and horticultural needs of the United States, but it was also a means to express in practical terms both his knowledge and his love of nature. In the garden he became a practical naturalist, putting his expertise to everyday use, which he greatly enjoyed. Writing to Charles Willson Peale in 1811, he mused: "I have often thought that if heaven had given me choice of my position & calling, it should have been on a rich spot of earth, well watered, and near a good market for the productions of the garden. No occupation is so delightful to me as the culture of the earth, & no culture comparable to that of the gardens. . . . But tho' an old man, I am but a young gardener."[46]

During the years 1766 until the autumn of 1824, two years before his death, Jefferson maintained a garden book, which began as a garden diary and grew to include all remotely garden-related topics that occurred to him. The entries ranged from plans for building roads and fish ponds, to observations on a great flood in Albemarle and comments concerning various horticultural results. Whatever he saw was recorded: the first blossoms of his flowers, moments to harvest the crops, information about the new plants he introduced, his attempts to improve farming or his experiences with viniculture.[47] In letters to numerous correspondents in the United States and Europe, he advanced his theories on agriculture and gardening, told what he was planting, requested information or plants from them, and ordered particular species or seeds from nurseries and seedsmen. One of the persons with whom he maintained an energetic conversation and exchange of plants and seeds was Madame de Tessé.[48] Beginning in 1784, when he was living in Paris, Jefferson also conducted an important horticultural exchange with André Thouin, the superintendent of the Jardin des Plantes in Paris.[49] In a letter written near the end of his life, he refers to this important, long-lasting relationship and "my good old friend Thouin . . . who for three and twenty years of the last twenty five has regularly sent me a box of seeds, of such exotics, as to us, as would suit our climate, and containing nothing indigenous to our country."[50] While he was away from Monticello, Jefferson sent his daughter Maria many letters containing gardening instructions and information, while describing to her the natural world around him at whatever place he happened to be staying. In a letter to Harry Innes about natural history and politics, Jefferson writes: "The first is my passion, and the last my duty, and therefore both desirable."[51]

Jefferson's fascination with natural history is strongly evident in his fierce battle with Buffon over the assumed inferiority of all species of the American continent. He devoted considerable effort over many years to the destruction of this theory: as a scientist he wanted to prove that Buffon was wrong, and as an American he wanted to defend his continent. Rather than merely pursuing a theoretical argument, Jefferson settled on a simple empirical test: a direct comparison of the sizes of mammals in Europe and America. Thus he began to systematically and scientifically dismantle Buffon's works in his *Notes on the State of Virginia.* When he arrived in Paris in 1785, Jefferson sent the skin of a large panther to the French scholar; Buffon responded by inviting him for dinner. "In my conversation with the Count de Buffon on subjects of Natural history," Jefferson writes to Archibald Cary, "I find him absolutely unacquainted with our Elk and our deer. He has hitherto believed that our deer never had horns more than a foot long; and has therefore classed them with the roe-buck, which I am sure you know them to be different from."[52] Jefferson followed up his conversation with Buffon by sending him a moose so large, Jefferson claimed, that a European reindeer could walk under it.[53] In the letter to Buffon that accompanied this offering, Jefferson refers—undoubtedly with some degree of triumph—to his theories in *Notes on the State of Virginia,* a copy of which he had earlier sent Buffon.[54] Unfortunately, Buffon died only six months after receiving Jefferson's gift, before he could make any corrections for the following volume of his *Histoire naturelle, générale et particulière.*

Paleontology also greatly interested Jefferson, and his early writings on that subject are considered to have initiated the science of vertebrate paleontology in the United States.[55] Marshalling evidence in *Notes on the State of Virginia* to disprove the theory of American inferiority, Jefferson mentions the mastodon[56]—which he refers to confusingly as a mammoth—in his list of superior American fauna, stating that the "bones of the Mammoth which have been found in America, are as large as those found in the old world."[57]

When, in 1796, Jefferson was sent fossils discovered in Greenbriar County, in today's West Virginia, he speculated that the bones discovered belonged to an animal that had not yet been described in the annals of science.[58] Assuming that these bones came from a giant, clawed, lion-like creature, he assigned it to the "family of the lion, tiger, panther etc." and gave it the name *Megalonyx,* or giant claw, arguing that the "animal must have been as preeminent over the lion, as the big buffalo was over the elephant. The bones are too extraordinary in themselves, and too victorious an evidence against the pretended degeneracy of animal nature in our continent."[59] In 1797, Jefferson presented a paper with his interpretations of this finding and tables of comparative measurements at the American

Philosophical Society. Two years later, the paper was published with the following conclusion: "We may safely say that he was more than three times as large as the lion: that he stood pre-eminently at the head of the column of clawed animals as the mammoth stood at that of the elephant, rhinoceros, and hippopotamus, and that he may have been as formidable an antagonist to the mammoth as the lion to the elephant."[60] In the same volume, however, Caspar Wistar published an article in which he correctly identified the remains as belonging to a giant ground sloth.[61]

Jefferson learned from an issue of *Monthly Magazine and British Register of London* that had been published in September 1796 that his classification of the skeleton as a member of the cat family had been wrong. Similar fossils had been found in Paraguay and sent in 1788 to the Royal Cabinet of Natural History in Madrid. There the skeleton was assembled and anatomically described by the Spanish naturalist Juan Bautista Bru y Ramón.[62] The article was an abbreviated translation of a longer publication by Georges Cuvier, who had classified this creation as *Megatherium* and identified it as a distant relative of the sloth. In 1804, Cuvier published a more complete account of the *Megalonyx* and the *Megatherium*, giving Jefferson full credit for the *Megalonyx*. In 1822, the animal was named *Megalonyx jeffersonii* in honor of Jefferson's efforts to classify it.[63] In later years, Jefferson continued to send specimens to Paris as a way of maintaining his place in the transatlantic debate on natural science, as, for example, when he made a gift of American fossils to the Institut National de France, which had elected him as a foreign associate in 1801.[64]

Over the years, Jefferson's Monticello had become a showcase for his interest in natural history. In its entrance hall, a public reception space turned into a veritable cabinet of curiosities, he displayed his extensive collection of fossils, antlers from elk and moose, and anthropological objects, many of them gathered by Lewis and Clark.[65] This collection was a personal museum, with souvenirs from exchanges Jefferson had made in diplomatic and social contexts, but it was also an illustrative answer to Buffon. Jefferson, like Humboldt, had abundant respect for the great French naturalist, which is another reason he found it so important to convince Buffon to modify his stance on America.

Jefferson was also vitally interested in the Indian population living in the world around him. His personal encounters with the native peoples began during his boyhood in Virginia among white farmers who were trying to obtain land from the native population, and extended through his public career, when as U.S. president he had to develop a political program for the Indians, purchasing land from them and trying to establish peace with the tribes of the Louisiana Territory into his retirement.[66] Again Jefferson is shown caught between his scientific interests, his moral convictions, and his political decisions. Two of his lifelong moral dilemmas were his attitudes toward the black and the Indian populations, since they presented a challenge to his ideals of liberty and freedom in the new world.

Jefferson defended the Indians, both in his *Notes on the State of Virginia* as well as in numerous letters. He wrote to Chastellux that "I am safe in affirming that the proofs of genius given by the Indians of N. America, place them on a level with Whites in the same uncultivated state. The north of Europe furnishes subjects enough for comparison with them, and for a proof of their equality. . . . I beleive [*sic*] the Indian then to be in body and mind equal to the whiteman."[67] This spirited defense of the native American population is unmistakably related to Jefferson's American-inferiority debate with Buffon. At the end of his *Notes on the State of Virginia,* he added the speech of the Mingo chief Logan, who mourned the loss of his family in an attack by a white settler, as well as other documents related to that case, presenting "Logan's Lament" as an example of outstanding and powerful oratory at the level of European speakers. "I may challenge the whole orations of Demosthenes and Cicero," Jefferson writes, "and of any more eminent orator, if Europe has furnished more eminent, to produce a single passage, superior to the speech of Logan, a Mingo chief, to Lord Dunmore, when governor of this state."[68]

In contrast to his conclusions regarding black people, Jefferson considered the assimilation of the Indian population feasible, so they represented a comparatively minor problem to him. Only their living conditions, he felt, needed to be changed in order to make them full Americans. Despite the scientific and political interest Jefferson demonstrated in the Indians, he paid little attention to them philosophically and certainly did not romanticize them. Apparently uninterested in the then-current European discussions of the "noble savage," Jefferson focused on where the Indians came from; the differences between the tribes and their languages; and the dissimilarities among the Indians, Europeans, and Asians beyond those that resulted from diverse circumstances.[69]

Similarly, Jefferson charged Lewis and Clark's Corps of Discovery with gathering detailed information about the native inhabitants of the regions they crossed, particularly concerning their languages and clothing:

> In all your intercourse with the natives, treat them in the most friendly & conciliatory manner which their own conduct will admit; allay all jealousies as to the object of your journey, satisfy them of it's innocence, make them acquainted with the position, extent character, peaceable & commercial dispositions of the US. of our wish to be neighborly, friendly, & useful to them, & of our dispositions to a commercial intercourse with them; confer with them on the points most convenient as mutual emporiums, and the articles of most desirable interchange for them & us. If a few of their influential chiefs within practicable distance, wish to visit us, arrange such a visit with them, and furnish them with authority to call on our officers, on their entering the US. to have them conveyed to this place at the public expence. If any of them should wish to have some of their young people brought up with us, & taught such arts as may be useful to them, we will receive, instruct & take care of them. Such a mission whether of influential chiefs or of young people would give some security to your own party.[70]

For this contact with the Indian native inhabitants, special silver medals with a portrait of Jefferson and a message of friendship and peace had been prepared. The message of the so-called Indian Peace Medals was twofold, however, since they also were meant to symbolize the sovereignty of the U.S. government over the indigenous inhabitants. Many of the examples of the Indians' material culture and the zoological specimens that Lewis and Clark collected later formed part of the museum that Charles Willson Peale had established in Philadelphia; others were incorporated in Jefferson's collection of decorative Indian items. As a scholar, Jefferson was interested in understanding the native American population in their original conditions, before being altered by the influence of the Europeans. This lifelong interest contributed materially to the development of the nascent disciplines of anthropology, ethnology, and comparative linguistics in the United States. By the end of his presidency, Jefferson had collected lists of vocabularies of some fifty different Indian languages, many gathered by Lewis and Clark and some others by himself during his "northern Journey" with James Madison in 1791.

As a politician, however, Jefferson saw only two choices for the Indians: assimilate or be destroyed. At first he had hopes that they could be pushed off to the West, and he showed a high regard for those who settled down as agrarians, but if they did not assimilate and join his nation of farmers, he stood ready to defeat them militarily. Impelled by the necessity of expanding his agrarian republic, he designed a policy of removal that led to cultural genocide. As Wallace notes, while Jefferson the naturalist compiled Indian vocabularies, chronicled the eloquence of America's native peoples, and bemoaned their tragic fate; Jefferson the imperialist was the architect of Indian removal.[71]

Despite the dissimilarities in Jefferson's and Humboldt's upbringing and circumstances—Jefferson as a Virginia countryman and Humboldt as a cosmopolitan urbanite—their approaches to the natural world have certain aspects in common. This can be partially explained by the fact that they were both men of the Enlightenment, but beyond that, it shows a sympathy of mind and interest that became the basis for their lifelong friendship.

Their approach to and understanding of nature was marked by the ideas predominant during the Age of Reason—they preferred what could be concretely established to speculation, and delighted in measuring all things with their beloved scientific instruments. Nonetheless, years before the Romantic era began, their letters and other writings reflected a romanticized attitude toward nature, and particularly the majesty of certain landscapes. Their enthusiasm for natural bridges offers one example. Humboldt was fascinated by the phenomenon, and in his work *Views of the Cordilleras* he dedicates a whole chapter to the Natural Bridge in Iconozo, which he had crossed on his way from Santa Fé de Bogotá to Popayán and Quito in September 1801.[72] In a detailed description of this bridge, accompanied by one of his drawings, he refers to the Natural Bridge in Rockbridge County and in particular to the comments Jefferson had made concerning it in his *Notes on the State of Virginia.*[73]

Both Jefferson and Humboldt were involved in the scientific debates of their time—the theory of American inferiority, the climate-effect controversy—and maintained an active interest in a number of different fields, advancing their knowledge by reading, by applying the knowledge they had gained from their own experiments, and by exchanging and comparing information, theories, and ideas with other learned men of their time. One thing that characterized both men's study of the natural sciences was the desire for a transatlantic exchange of perspective. Neither specialized in a particular scientific field; both were polymaths convinced of the interconnectedness of natural phenomena. Humboldt pointed out the connection of climate and soil formations with the distribution of plant and animal life, and the importance of geographic conditions to the development of mankind. Jefferson, in his *Notes on the State of Virginia,* connected the natural history of a region to its people and its government. He not only described rivers, but he outlined their relationship to commerce and trade. He classified the plants and trees as to their value for ornamental, medicinal, and esculent purposes and included a comparative view of America's native birds and animals with those of Europe.[74] He correlated the data he had accumulated with distinct phenomena, such as the impact that the elimination of forest had on the changes of climate or the importance of the observed the weather conditions. The difference is that Humboldt, as a scientist, was able not only to deepen his ideas and develop theories based on them, but also to expand his conclusions in a much broader way. Thus today it is Humboldt who is considered a forefather of modern environmental thinking, whereas Jefferson's early contribution to the foundation of what we today call ecology has received much less attention.

The affinities in their approach to nature are apparent in their published descriptions of particular regions with which they were familiar—Virginia in Jefferson's case and New Spain and Cuba in Humboldt's. These publications, not merely descriptive, but including economic and demographic information as well as addressing political and moral concerns, went beyond the scope of the works on natural history typical of the time. Humboldt called the works in this new category "Political Essays," and the fact that he owned a copy of Jefferson's *Notes on the State of Virginia* and referred to it respectfully suggests that Humboldt was inspired by the Virginian's method of describing a particular geographic region, taking into account its population, history, plants, animals, nature, climate, and mineral sources. Jefferson's work may have served as a model for the two regional essays he wrote after his expedition.[75] That Jefferson perceived the works' similarity can be seen in his comment, on sending Humboldt the long-requested signed copy of *Notes,* that his own effort must surely appear weak to the famous author of the excellent work on South America.[76]

8 Parallels and Discrepancies

This chapter considers three distinct yet connected topics intended to highlight the differences and similarities in Humboldt's and Jefferson's convictions and the ways that each man expressed his convictions through actions. The first part focuses on their studies related to geography and explores the extent to which their thinking was influenced by the German scholar Bernhard Varenius, considered to be the founder of scientific geography. The second part analyzes the ideas and interests Jefferson and Humboldt shared, as well as the values they understood and applied differently concerning the concept of liberty, the progress and dissemination of knowledge, the meaning and function of religion, and the concept of the nation. The third part establishes the connection of their views to the European and American versions of the Age of Reason.

Chain of Ideas: The Influence of Bernhard Varenius

Bernhard Varenius's *Geographia generalis*, published in 1650, is considered to be his most important work and one of the first to introduce a more systematic approach to the subject of geography; it set the scientific standard for more than a century. *Geographia generalis* was also the first textbook to be used in general geography courses in an American college.[1] Thus, it is interesting to note the extent to which the ideas, as well as the conception of geography expressed in Varenius's book, may have influenced two of his famous successors, Jefferson and Humboldt, and to what extent they may have further developed his ideas. The history of ideas is a continuum: theories are taken from scholars of the past and developed; one's own ideas are taken by future generations and molded by them.

The Greek term *geographia* means "to describe or write about the Earth" and is therefore a very broad concept that encompasses the study of the land, its features and inhabitants, and other phenomena. Eratosthenes (276–194 BC) was the first person to use the word "geography," which is considered to be among the oldest of all sciences and from which developed other scientific fields such as biology, anthropology, geology, mathematics, astronomy, and chemistry. In order to establish a division within geography, William Pattinson defined four historical traditions within it: spatial analysis of natural and human phenomena, area studies, study of the man-land relationship, and research in the earth sciences.[2] Despite its long history, however, it was not until the eighteenth and nineteenth centuries that geography was recognized as a formal academic discipline and became part of a typical university curriculum. Many geographic societies were formed during the nineteenth century, such as the Société de Géographie in 1821, the Royal Geographical Society of London in 1830, the Russian Geographical Society in 1845, the American Geographical Society in 1851, and the National Geographic Society in 1888. Modern geography is an all-encompassing discipline that still audaciously seeks to understand the earth and all of its human and natural complexities. It is commonly divided into two major branches: cultural or human geography, and physical geography. *Cultural geography* deals with the impact of human activities on the earth, while *physical geography* studies the features of the natural environment, in particular elements, processes, and patterns such as the atmosphere, hydrosphere, biosphere, and geosphere. From the birth of geography during the Greek classical period until the late nineteenth century, the field of geography was almost exclusively a natural science. While Humboldt and Carl Ritter are considered the founding fathers of modern geography, Humboldt's contributions are stronger in the field of the physical geography, while Ritter was more concerned with human geography. The connection between the physical and human properties of geography is most apparent in the theory of environmental determinism. Popularized in the nineteenth century by Ritter and later by Friedrich Ratzel, environmental determinism sees in physical, mental, and moral habits a direct connection to the influence of natural environment.

Both Humboldt and Ritter are recognized for their efforts to turn geography into a formal academic discipline: Humboldt's five-volume *Cosmos* is considered a pivotal work and a successful attempt to unify various branches of science and philosophy into a comprehensive treatment of the physical world. In 1828, Ritter established the Berlin Geographical Society (Gesellschaft für Erdkunde zu Berlin) and occupied the first chair in geography at the University of Berlin. Jefferson's case is different. In light of his numerous achievements as a politician and his contributions in other fields of sciences, relatively few scholars have dealt with him exclusively as a geographer.[3] For many years before he finally dispatched the Corps of Discovery to the Pacific, he had been especially interested in the scientific exploration of the unknown western regions of the North American continent. In fact, this geographical interest was the primary reason Jefferson first invited Humboldt to Washington, as is evident from both their initial and their continuing communications. Nevertheless, Jefferson's approach was different from Humboldt's: with the Lewis and Clark expedition as well as in his discussions with European visitors, his study of geography was more political and practical than academic. His primary aim was to map accurate borders and determine the commercial possibilities of rivers in connection with his larger plans for the American West.

Because of their respective importance to the development of geography, and likely also as a result of their personal contact, there are several ways to link Jefferson and Humboldt to Bernhard Varenius.[4] Both men appear to be, to a certain extent, explicitly or implicitly inspired by Varenius. Albert Bergh argues that Jefferson "may be said to stand between Bernhard Varenius, who in 'Geographia generalis,' 1650, essayed the interpretation of the climatic conditions and the physical changes of the earth's surface, and Humboldt's *Kosmos*, 1845. The latter supplemented Varenius by pointing out the connection of climate and soil formations with the distribution of plant and animal life, and even more importantly the relation of geographic environment to the development of mankind, especially as to colonization, commerce and industry. Jefferson's *Notes on Virginia*, fifty years in advance of Humboldt, presaged the latter's advances scientific geography."[5]

The first step in an analysis of Varenius's possible influence on Jefferson and Humboldt, or at least in the establishment of conceptual similarities in their works, consists of searching their works for direct references to Varenius. With regard to Jefferson, this research produced no positive evidence. Specifically, consulting the different editions of Jefferson's papers and their indexes revealed no reference to the German geographer.[6] These results correspond with an analysis of all the references to previous works that could be found in Jefferson's *Notes on the State of Virginia.* In this work, Jefferson demonstrates that he had conducted major research on geographical questions, and that he was acquainted with numerous works from all over the world in different languages, which he quotes, mentions, or comments upon. Among the authorities cited in the text and footnotes, not a single mention of Varenius could be found. Finally, the catalogue of books purchased from Jefferson by the Library of Congress was consulted to see if he owned *Geographia generalis.* Once again, in a very large list of books on geography,[7] this work did not appear.

The situation was not much different in regard to Humboldt: out of all his works, only the first volume of the final synthesis of his lifelong studies, the five-volume *Cosmos,* contains references to Varenius. Nonetheless, one lengthy mention in the chapter "Physical Description of the World" is quite interesting. Humboldt there writes of Varenius that he

> was the first to distinguish between general and special geography, the former of which he subdivides into an absolute, or, properly speaking, terrestrial part, and a relative or planetary portion, according to the mode of considering our planet either with reference to its surface in its different zones, or to its relations to the sun and moon. It redoubts to the glory of Varenius that his work on General and Comparative Geography should in so high a degree have arrested the attention of Newton. The imperfect state of many of the auxiliary sciences from which this writer was obliged to draw his materials prevented his work from corresponding to the greatness of the design, and it was reserved for the present age, and for my own country, to see the delineation of

comparative geography, drawn in its full extent, and in all its relations with the history of man, by the skillful hand of Carl Ritter.[8]

Thus, after a short explanation of Varenius's basic concept and a reference to the importance of Newton in the diffusion of these ideas, Humboldt asserts that his predecessor had laid the foundation for a new focus on geography, though it could not be performed comprehensively during his time. Clearly he saw his own work as a continuation of that of Varenius.

In a long footnote, Humboldt adds to this comment a more detailed view of the importance and the limits he ascribed to Varenius's work, starting with some general reflections on *Geographia generalis:* "This excellent work by Varenius is, in the true sense of the words, a physical description of the earth. Since the work *Historia natural y moral de las Indias,* 1590, in which the Jesuit José de Acosta sketched in so masterly a manner the delineation of the New Continent, questions relating to the physical history of the earth have never been considered with such admirable generality. Acosta is richer in original observations, while Varenius embraces a wider circle of ideas, since his sojourn in Holland, which was at that period the center of vast commercial relations, had brought him in contact with a great number of well-informed travelers."[9] He considers Varenius alongside the famous Spanish missionary and naturalist José de Acosta (1540–1600), whom he describes on several occasions as his intellectual predecessor in physical description of the earth.[10] Of Varenius, he maintains that no one since Acosta had written as good and general a description of the telluric phenomena of the earth.

Jefferson, too, uses Acosta as a point of orientation and refers to the Spanish Jesuit several times in his *Notes on the State of Virginia.* We know that Jefferson possessed his own copy of Acosta's famous work *Historia natural y moral de las Indias.*[11] Although these references may not be rich in content, they do demonstrate Jefferson's interest in Acosta. It seems curious that both Jefferson and Humboldt appear to have valued Acosta's work more than Varenius's *Geographia generalis.*[12]

Besides Humboldt's long, explicit reference to Varenius, implicit parallels between their concepts of geography can be detected. To a lesser degree, this is also the case with Jefferson. As noted in chapter 7, from the very beginning of his scientific activities, Humboldt clearly articulated the aim of his research and his holistic vision of the earth as an intricately interlocking organic whole. This interdependency resonates with the ideas of Varenius, who also saw a basic unity in a single earth-wide system, with close linkages among all terrestrial, celestial, physical, and human forces. As Varenius lamented, geography was criticized in his time as being either too narrowly descriptive or too broadly expansive, since readers were bored by a bare enumeration and description of regions without an explanation of the customs of the people.

Concerning Jefferson, we see the scientific curiosity of an Enlightenment scientist and researcher combined with a pragmatic orientation toward the whole cosmos. His approach to geography, in the final analysis, reflects what Varenius calls *special geography,* which is the description of particular places, versus *general geography,* which denotes the study of general and universal laws or principles that apply to all places. Varenius insisted on the practical importance of the type of knowledge included in the category of *special geography,* which suited Jefferson in his function as a representative of the American Enlightenment. Although he studied the abstract concepts or universal laws characteristic of general geography, he was not interested in elaborating upon them. His specific geographical interest derived from his work as politician.

By contrast, Humboldt's two regional studies—his essays about Cuba and New Spain—were not limited to physical descriptions, but delved into principles of general geography as well. In his last work, *Cosmos,* he elaborated on these theoretical aspects of geography. Thus, like Varenius, Humboldt is regarded as making his outstanding contribution more to the organization of information than to any particular field of study.

It is in this connection that the attitude of the three scholars regarding the position of man within nature becomes clear. In Varenius's concept of *special geography,* the "human properties of a place" form a third category, in addition to celestial and terrestrial properties. He considered cultures, language, government, and religion as elements that ought to be examined alongside factors like climate, surface features, minerals, animals, and plants. Humboldt included human beings in his concept of *physical geography*, considering them an integral part of nature; he did not, however, take humans as primary determinants, or assign them a special place in his research. With Jefferson, it was very much the same.

The three men were also united in their comparative approach to science. In the concept of plant geography he developed, Humboldt recognized the interdependence of areal incidences and the need for explaining any one set of regionally distributed phenomena in relation to its spatial context. Therefore, he repeatedly compared areas with similar landscapes in different parts of the world. In this, his thinking obviously parallels that of Varenius, who divides his *general geography* into absolute, respective, and comparative parts that in turn arise from the connection of diverse places on the earth. In Jefferson's case, too, this comparative method is manifested in his refutation of Buffon's ideas on the supposed inferiority of America.

A final similarity in the work of the three authors can be observed in their attitudes toward the role of theology in science. Although geography in Varenius's time was closely linked to religious questions, he dissented from that view, and his work can be considered the first scientific approach to the discipline.[13] Humboldt and Jefferson were in accord with this thinking, despite the problems they individually faced as a result: the Prussian in trying to disconnect theology from science, and Jefferson in establishing a new secular form of education that culminated in the creation of the University of Virginia.

The obvious parallels in the contributions of Varenius, Humboldt, and Jefferson to the systematization of geographical knowledge available during their time may not be the result of a direct transmission of ideas. They integrated previous conceptions, in some cases derived from the classical Greeks and Romans, and enriched contemporary geographical understanding with their own observations and conclusions. Nevertheless, what can be shown is the general process of inspiration through other works—through the implicit elaboration of similar intellectual models or explicit references—in brief, the evolution and continuation of the *chain of ideas*. If we look at the larger picture, we see that Varenius, too, absorbed the ideas of previous and contemporary thinkers, particularly Bartholomäus Keckermann,[14] just as in several of Humboldt's texts, references to Jefferson can be found, and in Charles Darwin's writings as well as in his private letters, he repeatedly states that he was very much inspired by the work of Humboldt, whom he calls "the greatest scientific traveler who has ever lived."[15]

Interpretations of Enlightened Values

Analyzing the thoughts and actions of Humboldt and Jefferson brings to light both similarities and significant differences between these two enlightened personalities. Besides those discussed in the previous chapters, other interesting affinities and discrepancies merit mention.

First, it is obvious that liberty in all its facets was a pivotal value for both of them. On a personal level, Humboldt tried to maintain his professional independence and detachment as long as possible. Politically, he showed the value he placed on liberty on countless occasions when he praised the free nations of America in contrast to colonial societies. Jefferson, for his part, repeatedly stated the importance he attached to personal and economic, as well as political, liberty. A 1791 reference to the conflict of interest between the federal government and the states shows that he was conscious of the difficulties besetting the American experiment in freedom: "I would rather be exposed to the inconveniences attending too much liberty than those attending too small a degree of it."[16] For both Humboldt and Jefferson, the happiness and prosperity of the people was always foremost in their minds and directed many of their actions. They saw the wealth of nations in terms of the well-being of individuals, which they considered as the cornerstone of a progressive society.

Both men worked toward the dissemination of knowledge as another basic value of Enlightenment. Jefferson was interested in all branches of science, including geography, geology, astronomy, botany, zoology, medicine, agriculture, and chemistry, and always sought to glean from these fields information that would be useful for the people. He recognized the value of and the need for such information, and had a sense of the important role it should play in the future of the American nation.[17] In *Notes on the State of Virginia,* he points out the importance of agricultural education and research as well as the value of farm societies, given the relevance that agriculture had for the young nation. In his later presidential years, his government took the progress of science as a serious goal. Jefferson's effort to further scientific advances as well as to popularize knowledge is also demonstrated through his activities for the American Philosophical Society in Philadelphia, and later in his founding of the University of Virginia, where the curriculum was based upon his declared convictions.[18] Humboldt, for his part, not only published a great many scientific books and articles about his American expedition (and later a few on his Russian expedition), but also delivered his famous *Kosmos* lectures in 1827–28 in Berlin, with which he particularly hoped to reach a broader public.[19]

Both Humboldt and Jefferson were actively involved in establishing a network of scientists, politicians, artists, writers, and others from the Old and the New World dedicated to the exchange of information, ideas, and opinions. The exploration of America was another mutual interest, promoted by Jefferson with the Lewis and Clark expedition during his presidency, and by Humboldt not only through his own explorations, but also through the various kinds of support—intellectual, financial, and by letters of introduction—he provided to a large number of European travelers to America.

Another topic of common purpose was their formulation of a response to the debate on the assumed inferiority of America. Obviously, different motivations underlay their refutations of these inferiority theories. For Jefferson, certainly, there was an element of pride as an American and a man of science capable of correcting important misconceptions being published by European scientists. Unlike Jefferson, Humboldt was not personally invested in this debate; he responded on an academic—specifically scientific—level. Though Humboldt respected Buffon as a naturalist, he nevertheless criticized the myth Buffon had concocted, stating that such ideas could easily be propagated because they "flattered the vanity of Europeans." As an enlightened man of science who sought to understand the world by taking measurements, he refuted the assumed inferiority of America on the basis of his own scientific results. Despite their different motivations for doing so, Jefferson and Humboldt participated on the same side of the transatlantic debate about Buffon's theory.

Further parallels can be detected with regard to the meaning and function of religion. Both men considered religion, especially Catholic doctrine, as an obstacle to the development of science, and on several occasions complained of the negative impact of religious dogmatists on society, whether in the Old or New World. Jefferson had been a relentless rationalist since his youth, and he viewed the principles of Christianity with great skepticism.[20] In this sense, his self-appointed task to extract from the Bible the authentic sentiments of Jesus has to be understood as an attempt to get at the essence of Christianity. Jefferson's interest in philosophy and science conflicted with traditional religious dogma, causing him many difficulties. During the course of his political career, charges of atheism were often leveled against him, particularly by his Federalist opponents during the heated presidential campaign of 1800. This explains Jefferson's reticence in voicing his personal convictions and making public pronouncements concerning religion. Nevertheless, he apparently had enough confidence in Humboldt to reveal his concern about the role of religion in the future of Latin America.[21] Humboldt himself was attacked in clerical circles for not following a traditional religious doctrine and for criticizing the church and the Christian mission among the Indians in America. The details of this controversy even reached the United States, causing people to take sides either for or against him.[22] Even in the context of the immense celebration of the centennial of Humboldt's birth in 1869, this topic—the lack of God in his work—was debated in newspapers as well as in speeches in the United States. Although his science was widely accepted and followed, the philosophy on which his scientific achievements were based proved problematic. There seemed to be difficulties in accepting an approach to natural science developed in isolation from religious teachings.[23]

There were also parallels in Humboldt's and Jefferson's ways of life. As young men, both inherited fortunes that enabled them to carry out their own plans, and each sacrificed his own fortune for the achievement of his visions and ideas. As a consequence, at the end of their long lives they faced serious financial difficulties. Jefferson sold his extensive library to Congress in 1815, and in 1829, after his death, his family was forced to auction his slaves, many of the family's possessions, his remaining books, and even his home, Monticello. Humboldt, for his part, exhausted his funds on his American exploration and in living for many years afterward in Paris, working on the publications resulting from his expedition. Financial need forced him to return to Berlin in 1827 to work in the service of the Prussian king.

Despite these many parallels, however, and similar approaches to certain issues, there were also several differences between Jefferson and Humboldt, particularly in regard to science. Jefferson viewed scientific advances in terms of their practicality in the building of a new society. He was primarily interested in the invention or adaptation of new and useful instruments, such as the mould-board plow,[24] for which he was awarded a gold medal by the Agricultural Society in Paris; the "wheel cypher," a cryptographic device that employed concentric rings to scramble or unscramble letters in a secret message; and other creations such as a portable music stand for a quintet and a revolving chair. His reputation as an inventor was derived primarily from his compulsion to modify and attempt to improve existing utilitarian objects and devices to meet his own requirements, many of which he designed for his own comfort at Monticello.[25] As he wrote to Benjamin Waterhouse: "The fact is, that one new idea leads to another . . . until some one, with whom no one of these ideas was original, combines all together, and produces what is justly called a new invention."[26]

Humboldt was also interested in the applied sciences and their uses toward the improvement of living and working conditions. For example, he developed several inventions useful to miners, including special lamps and other mining-safety devices. His approach to science, though, was also very theoretical, based on the acquisition of knowledge solely for the sake of broadening or deepening the understanding of nature. He took part in theoretical controversies such as, for example, the late-eighteenth- to early-nineteenth-century Neptunism versus Plutonism debate regarding the origin of the rocks comprising the earth's crust. While the Neptunists—led by Abraham Gottlob Werner, Humboldt's professor at the Mining

Academy of Freiberg in Saxony—believed that all rocks had been formed in water, the Plutonists argued that the earth was formed in fire by *volcanic activity*, with a gradual, ongoing process of weathering and erosion wearing away rocks.

Both Humboldt and Jefferson had an interest in pure scientific progress as well as in the improvement of society brought about by implementing those scientific achievements, but whereas Humboldt dedicated his life entirely to his main goal, the pursuit of scientific knowledge, Jefferson focused more on politics, often leaving his scientific projects aside or appointing others to carry out his tasks. These differing priorities relate directly to the historical contexts in which Humboldt and Jefferson lived and the demands of their respective circumstances. They also relate to their personalities—Humboldt was more academic; Jefferson more interested in political power. Jefferson often maintained that he would have liked to stay away from politics, but circumstances—the creation of a new nation in which he felt obliged to actively take part—would not allow it.

Another important difference was their attitudes toward their nationality. While Jefferson felt close ties to his country throughout his life, Humboldt always showed that his aspirations and ideas were not limited by the boundaries of Prussia, the German states, or even Europe. As Francis Lieber stated in his speech at the unveiling of the Humboldt statue in New York on September 14, 1869: "Humboldt, though a German in his lineaments of character and talents, was of all the modern men the one whose endeavors, aspirations, and fame were least limited by national demarcations."[27] Conversely, Jefferson clearly expresses his self-identification as an American in one letter to Humboldt, telling him: "I am held by the cords of love to my family & country, or I should certainly join you."[28]

There was also a very sharp distinction between their actions and positions regarding slavery. Humboldt was free to express his aversion to the institution openly: Jefferson, on the other hand, had to be mindful of many considerations other than his own personal convictions. In addition, Humboldt had the luxury of evaluating the institution from outside, having never been connected to slavery in any way. Jefferson, however, was born into a slaveholding family, and he was inextricably linked to that long-established practice among Virginia planters. Slaves formed a significant part of his personal wealth and provided him with the income and liberty to dedicate his time to the things that were important to him. The decision to eliminate slavery would have had major consequences for his own political and personal life.

The same explanation can be applied to Jefferson's political views on Indians. His paternalistic attitude does not coincide with Humboldt's views on the topic.[29] The latter expressed deep concern over the oppression of the native population and slaves, whose liberation he advocated vigorously on all occasions. His attitude toward the Indians was markedly influenced by his humanist ideals, nurtured by the exotic travel literature he read in his youth, and sealed in his contact with them in the New World, where he had seen them mostly in a deplorable state of oppression. Jefferson's view, on the contrary, was based on his activity as a politician. He was personally confronted with this issue as a political problem that required practical solutions. Therefore, whereas for Humboldt the Indians were part of a faraway, exotic reality, for Jefferson they formed a part of his daily political life. And as ever, his official statements may not have represented his personal convictions.

The European versus the American Enlightenment

The postulates of the Age of Reason grew simultaneously on both sides of the Atlantic in a cross-fertilization of ideas and ideologies. The Enlightenment was directed to the improvement of the structure of society and human life in general, which was considered a common task between the New and the Old Worlds. Ignorance was the enemy, and believed to be the major obstacle to progress. There was also mutual stimulation through literature in general, and travel narratives in particular. Nevertheless, Enlightenment ideas found very different expression, and led to different consequences, among the European countries; between the Old and the New Worlds such differences were even more pronounced.[30]

In Europe, there was usually a distinction between the philosophers and those governing. The philosophers had to argue against the traditional political practices and to establish the ideological bases for a new form of society. In America, the philosophers themselves had the chance to create the desired societal order based on their values. The dissimilarities between the American and the European Age of Reason have generated a great deal of interest and debate in international history.[31]

The Age of Reason postulates were not generic and cannot be understood as the products of a homogeneous movement; rather, they were shaped by the needs of a particular society at a particular moment. The numerous forms in which these convictions were manifested represented alternative approaches to modernity, alternative habits of mind and heart, of consciousness and sensibility.

The political and social ideas of the European version of Enlightenment have had a particular importance in American history. More universally accepted in eighteenth-century America than on the other side of the Atlantic, they were more widely incorporated into the formal arrangements of state and society, and they have lived on more vigorously into later periods. The peculiar force of these ideas in the New World resulted from various factors, many originating in the pre-Revolutionary period and the political experiences of the American colonists.[32]

Thus here we close the circle and return to our initial arguments, where we introduced Jefferson and Humboldt as representatives of the ideas of the Age of Reason, while acknowledging that the postulates, values, and priorities of the Enlightenment differed in the Old and New Worlds.

Many Europeans, for example, still preferred constitutional monarchies and thought the best path to a society's happiness and well-being was through a philosopher-king. Americans believed, however, that a representative republic would most safely and effectively provide for the general welfare, and that no other government could be reconciled with the fundamental principles of the Revolution. Jefferson's and Humboldt's views reflect this distinction. Humboldt performed his scientific projects in the frame of the Spanish monarchy, with which he stayed in contact after his return to Europe, dedicating his geographical essay on New Spain to King Carlos IV of Spain. After returning to Berlin in 1827, Humboldt worked for the Prussian court, and in spite of his political ideas, he did not criticize monarchy or courtly life as such. Jefferson, on the contrary, disliked monarchy in every form, as he often emphasized in his letters. He believed in self-government as a fundamental principle, though he accepted constitutional government as a way station in the progress toward a republic.[33] Politically as well as socially, he wanted the United States to distance itself from monarchy and everything related to it.

Another area in which the differences between Humboldt's and Jefferson's understanding of the ideals of Enlightenment reflect the differences between the Old and New Worlds is in their attitude toward science. The more pragmatic use of knowledge and its application to the improvement of the living condition of mankind in the United States differs from the more European Enlightenment view that the measuring and understanding of the entire world constitutes an end in itself. As we have seen, Humboldt played a larger role in creating, organizing, and interpreting new data, while Jefferson worked within political and social institutions to make use of the obtained information.

Finally, crucial differences exist in Jefferson's and Humboldt's views of the humanist ideals of the eighteenth century and the principle of equality among mankind. Jefferson's application of these values proved rather limited, given his views on slavery and his political decisions concerning the Indian population. In contrast, Humboldt applied these enlightened postulates to all people, reflecting the view of equality among humans that was key to European Enlightenment philosophy.

Epilogue

Alexander von Humboldt and Thomas Jefferson were both decisively marked by their transatlantic experiences. With their cosmopolitan worldviews, they were able to cultivate a dialogue between the Old Continent and the New World that had a positive impact on both sides. Their relationship, which reflected a strong personal affinity, was characterized by a mutual interest in science and politics.

An examination of the transatlantic contact and communication between Humboldt and Jefferson as the Enlightenment came to an end, as well as of their respective views on the events of their time, offers insight into the development of political thought and the progress of science. The exchange of knowledge and ideas between these two global thinkers also serves as an early demonstration of the importance of transatlantic communication and scientific cooperation.

Humboldt and Jefferson clearly remain relevant today. Each has been the subject of many books, as well as several exhibitions, documentaries, and movies. The Alexander von Humboldt Research Center at the Academy of Sciences in Berlin and the Robert H. Smith International Center for Jefferson Studies in Charlottesville, Virginia, were established to serve as focal points for the scholarly community engaged in analyzing and expanding the two men's legacies.

Humboldt is today considered a key figure in the study of the globalization of science. His well-organized and fruitful intra-European and transatlantic communication established new standards for intellectual networking. That a research institution dedicated to the study of the interdependence of the Internet and society was recently named after him attests to Humboldt's continued relevance in the scholarly community. Similarly, the Humboldt

Forum in Berlin, which aims to promote intercultural exchange and the exchange of global ideas, and to offer a multilingual meeting place for people with different cultural backgrounds, not only bears Humboldt's name but also seeks to honor and perpetuate his spirit. He was a great synthesizer of knowledge across disciplines, and his transcultural and comparative perspective and convictions continue to inspire scientific research and philosophical debate into the twenty-first century.

Jefferson's legacy is based on the pivotal role he played as one of the Founding Fathers of a new nation. As U.S. president, he contributed decisively to the establishment of what is today considered the American identity. His contributions are not only political but also include his scientific or technological work, his attempts to enrich the cultivation of vegetables and livestock, his writings about individual freedom, and the philosophy of higher education expressed in his creation of the University of Virginia. The legacy of Jefferson's engagement with aspects of European culture such as French cuisine, Italian and classical Greek architecture, and the art of viticulture is evident in U.S. culture today. Especially significant is Jefferson's establishment of a broad national and transatlantic network for the exchange of knowledge and ideas. In the spirit of his work, his lifelong interest in acquiring new knowledge, his approach to science and politics, and his creativity and persistence in carrying out his projects, Jefferson still serves as a model of active scholarly and civic engagement in society.

This analysis of the transatlantic communication between Humboldt and Jefferson delineates the foundations of twenty-first-century scientific networks and demonstrates how interconnected the sciences had already become in the late eighteenth and early nineteenth centuries. It also illustrates the impact of Humboldtian science—and, through Humboldt, European science—on the establishment of modern American science. Of more than historical interest, the analysis of the transatlantic scientific network during this time also raises questions about the value of scientific study for mankind and about how to disseminate and apply research findings in a modern, globalized world.

Appendix

Humboldt and Jefferson's Correspondence

ALEXANDER VON HUMBOLDT TO THOMAS JEFFERSON, PHILADELPHIA, MAY 24, 1804

Mr. President,[1]
Having arrived from Mexico to the blessed land of this republic, whose executive powers have been entrusted to your enlightened hands, it is my pleasant duty to present my respects and to express my great admiration for your writings, your actions, and the liberalism of your ideas, which have inspired me from my earliest youth. I had dearly hoped to be able to express my sentiments to you in person, at the same time delivering the enclosed package, which my friend, the Consul of the United States in Havana, has kindly requested me to give to you. Since the unloading of my herbarium has detained me here, and delayed my departure for Washington, I have been obliged to choose to use the postal service. The dreadful storm which was felt in Georgia made our voyage as dangerous as it was long (24 days), and I regret that the package has been so long in reaching you.

My desire to be of use to the physical sciences and to study mankind in its different states of barbarism and culture inspired me in 1799 to undertake, at my own expense, an expedition to the tropics. Thanks to a combination of fortunate circumstances and personal trust, the Spanish Government has granted me privileges beyond those enjoyed by La Condamine and the Abbé Chappe. I found in my friend Citoyen Bonpland, a student of the Paris Museum, great levels of knowledge, courage, and enthusiasm that should inspire all those who, through great sacrifice, strive for moral insight. For the past five years we have traveled throughout New Andalusia, the Carib and Chaimas Indian Missions, the provinces of Barcelona, Caracas, Varinas, and all of Guyana. We have covered almost a thousand nautical leagues by canoe on the Orinoco, the Guaviare and the Rio Negro, passing the huge and beautiful rapids of the Maipures and Atures on two occasions, and we determined by our chronometers, by longitudes and by the satellites [of Jupiter] the exact position of the Casiquiare, a tributary of the Orinoco which connects with the Amazon and by which we advanced to the borders of Grão-Pará [Brazil]. There in the wilderness and ancient forests of the Casiquiare, at Latitude 2° N, we saw rocks covered with hieroglyphs which prove that this remote land, now populated by a few naked Indian cannibals, was at one remote period the home of civilized peoples. Upon returning from the Rio Negro to Cumana we proceeded to the island of Cuba, and thence to the Sinú River, and Santa Fé [Bogotá], we traversed the Kingdom of New Granada, Popayán, and Pasto. For a year we carried out our work in the Andes of Quito, carrying our instruments up Mount Chimborazo to a height of 3,036 toises [19,413 feet], and thus 500 toises [3,197 feet] higher than any other human being before us. In order to study the chinchona trees we proceeded to Loxa, to the province of Jaén and onward to the Amazon. In Lima we observed the transit of Mercury, and sailing from there via Guayaquil for Acapulco we traveled for a year through the kingdom of New Spain, which offered us a vast field for observations. Despite the dangers of the climate for a young man, born in the frosts of Ultima Thule—Prussia—and despite the lack of food and shelter, to which I was exposed for several months, my health did not fail even for a single day. In spite of my burning desire to return to Paris, where I had long worked with Citoyens Vauquelin and Chaptal, and where we now hope to publish our works (the fruits of this expedition), I could not resist the moral interest to see the United States, and to enjoy the comforting aspect of a people which understands the precious gift of Liberty. I wish it were possible for me to present my personal respects and admiration to you, and to admire in you a philosopher magistrate, who has gained the approval of two continents!

Pray forgive, Mr. President, the confidential tone and length of this letter. I do not know whether my name is already known to you through my work on galvanism, or through my publications in the *Mémoires de l'Institut National de Paris*. As a friend of science, you will accept my expressions of admiration with indulgence. I should like to talk to you more about a subject that you so astutely dealt with in your work on Virginia, some mammoth teeth that we discovered in the Andes of the Southern Hemisphere at 1,700 toises [10,870 feet] above the level of the Pacific Ocean. My friend Citoyen Cuvier will give an anatomical description. It would be an abuse of your kindness to take up more of your time, and I shall confine myself to sending you the assurance of the deepest respect with which I have the honor to be,

Mr. President,
Your very humble and obedient Servant,
Baron Humboldt of the Berlin Academy of Science,
Philadelphia, May 24, 1804
Should you wish to honor me with your orders, please be so kind as to address them to me, care of Governor M'Kean.

THOMAS JEFFERSON TO ALEXANDER VON HUMBOLDT, WASHINGTON, MAY 28, 1804

I recieved last night your favor of the 24.th and offer you my congratulations on your arrival here in good health after a tour in the course of which you have been exposed to so many hardships and hazards. the countries you have visited are of those least known, and most interesting, and a lively desire will be felt generally to recieve the information you will be able to give. no one will feel it more strongly than myself because no one perhaps views this new world with more partial hopes of it's exhibiting an ameliorated state of the human condition. in the new position in which the seat of our government is fixed we have nothing curious to attract the observation of a traveller and can only substitute in it's place the welcome with which we should recieve your visit, should you find it convenient to add so much to your journey. accept I pray you my respectful salutations and assurances of great respect and consideration.

M. Le Baron de Humboldt.

THOMAS JEFFERSON TO ALEXANDER VON HUMBOLDT, WASHINGTON, JUNE 9, 1804

Thos: Jefferson asks leave to observe to Baron de Humboldt that the question of limits of Louisiana between Spain & the US is this. they claim to hold to the river Mexicana or Sabine & from the head of that Northwardly along the heads of the waters of the Mississipi to the head of the Red river & so on. we claim to the North river from it's mouth to the source either of it's Eastern or Western branch, thence to the head of the Red river & so on. Can the Baron inform me what population may be between those lines of white, red or black people? and whether any & what mines are within them? the information will be thankfully recieved. he tenders him his respectful salutations.

ALEXANDER VON HUMBOLDT TO THOMAS JEFFERSON, PHILADELPHIA, JUNE 27, 1804

Mr. President,
I told you in my first letter that I came to this country in order to see you, and to express personally the feelings of admiration and respectful affection that your writings, your ideas, and your actions have inspired in me for so many years. My departure tomorrow indicates that I have achieved the purpose of my visit. I have had the good fortune to see the first Magistrate of this great republic living with the simplicity of a philosopher, and receiving me with that profound kindness that can never be forgotten. My circumstances oblige me to leave, but I take with me the consolation that, while Europe presents an immoral and melancholy spectacle, the people of this continent are advancing with great strides towards the perfection of social conditions. I should like to think that one day I shall again enjoy this consoling experience, and I share your hope (which you expressed in the letter that Mr. M'Kean has kindly forwarded to me) that mankind may look forward to great improvements which can be expected from the new order of things to be found here. Please accept the expression of my highest esteem and respect, with which I shall for the rest of my life remain,

Mr. President,
Your very obedient and humble servant,
Humboldt
Philadelphia, June 27, 1804
My friends Bonpland and Montufar
send their respects to Your Excellency.

ALEXANDER VON HUMBOLDT TO THOMAS JEFFERSON, PARIS, MAY 30, 1808

Mr. President,

In the midst of all the misfortunes to which my country has succumbed, I have attempted from time to time to express the feelings of gratitude and admiration which I have for you. I fear that my last letter, dispatched by ship from Bremen, did not reach you. Today I have chosen a safer way to send you these lines, and to present you with my astronomical work and my *Essai Politique sur le Royaume de la Nouvelle Espagne.* In these works you will find your name cited, with that enthusiasm that it has always inspired in the friends of Mankind. May providence preserve you for the benefit of this world, towards which I strive to direct my labors and hopes.

I am, Mr. President, with the profoundest respect,
Your Excellency's
Very humble and most obedient servant
Humboldt
Paris, École Polytechnique, 30 May, 1808,
Montagne Ste. Geneviève
Please be so kind as to remember me to
Mr Madison and Mr Gallatin,
who honored me with their kindness.

THOMAS JEFFERSON TO ALEXANDER VON HUMBOLDT, WASHINGTON, MARCH 6, 1809

Dear Sir

I recieved safely your letter of May 30. & with it your astronomical work & Political essay on the kingdom of New Spain, for which I return you my sincere thanks. I had before heard that this work had begun to appear, & the specimen I have recieved proves that it will not disappoint the expectations of the learned. besides making known to us one of the most singular & interesting countries on the globe, one almost locked up from the knolege of man hitherto, precious addition will be made to our stock of physical science, in many of it's parts. we shall bear to you therefore the honorable testimony that you have deserved well of the republic of letters.

You mention that you had before written other letters to me. be assured I have never recieved a single one, or I should not have failed to make my acknolegements of it. indeed I have not waited for that, but for the certain information, which I had not, of the place where you might be. your letter of May 30. first gave me that information. you have wisely located yourself in the focus of the science of Europe. I am held by the cords of love to my family & country, or I should certainly join you. within a few days I shall now bury myself in the groves of Monticello, & become a mere spectator of the passing events. on politics I will say nothing, because I would not implicate you by addressing to you the republican ideas of America, deemed horrible heresies by the royalism of Europe. you will know, before this reaches you, that mr Madison is my successor. this ensures to us a wise & honest administration. I salute you with sincere friendship & respect.

Th: Jefferson

M. le Baron Humboldt.

ALEXANDER VON HUMBOLDT TO THOMAS JEFFERSON, PARIS, JUNE 12, 1809

Sir,

You are sufficiently aware of the respectful and friendly feelings that I bear towards you, to appreciate the satisfaction I felt at the receipt of your letter of March 6. I have not been happy since I left your wonderful country. When battered by storms, one becomes more sensitive to true moral pleasures. But what an amazing career yours has been! What a wonderful example you have given of energetic character, of graciousness and depth of the tenderest affections of the soul, of moderation and equanimity as the first magistrate of a powerful nation! What you have created, you see bearing fruit. Your retreat to Monticello is an event the memory of which will live forever in the annals of Mankind. It is difficult to talk to you about yourself without appearing to flatter you. No such artifice could be further from my open and emotional soul!

I present you with my works. I have the honor to present you with the second and third part of my work on Mexico, and the second, third and fourth parts of my astronomical anthology, including the surveys in the Andes.

I also enclose the translation that has been made of my *Tableaux de la Nature*, a translation which seems to be well done in English. Should the work be ready for shipment tomorrow, I shall also send the volume of our humble Society of Arceuil, wherein you will find my report on the respiration of fishes, and which presents the fine articles of my two closest friends, Gay-Lussac and Thenard. Do me the pleasure of accepting these trifles with your customary indulgence, with which you particularly favored me. I flatter myself that you may be pleased by my piece about the moral state of the Mexican people. I very much regret what I said on page ten about slaves. I have since learnt that when these lines were printed, Congress had already taken energetic steps toward total abolition. I was carried away by my devotion to the cause of the blacks, for which I have no need to blush. I shall make good the injustice I have committed regarding the Southern states in a note and in a supplement which will appear at the end of the work. My book was dedicated to King Carlos IV so as to pacify the displeasure the Madrid government might show to certain individuals in Mexico who furnished me with more information than the court might have wished.

I am sorry to learn that my letter of May 30 is the first to have reached you. Did you therefore not receive my work on the *Geography of Plants?*

Now I have a request to make of you. We are already 1,200 leagues apart. If I set off next year for Kashmir or Lhasa, I shall be even further away.

I possess your excellent work on Virginia, but should like to receive it from your own hands, with a line of your handwriting. It would be a very precious memento. You made me a gift of your own copy of Playfair, but your name is not in it, and I am afraid of this public misery, which divides everything into lines of red or blue. Please do not decline my request. Madame de Tessé, who is as devoted to you as I, says that my request is quite reasonable.

I do not dare to write to Mr. Madison, as I ought to have done. Let me congratulate the nation for the choice the American citizens have made. He made an excellent impression on me. I like your words "this ensures to us a wise and honest administration." The word 'honest' includes all that is just, liberal, and virtuous. If you write to the President, kindly send him the expression of my sincerest respect.

Be assured, my dear Sir, of my admiration and gratitude.
Alexander Humboldt
Paris, École polytechnique, Montagne Saint Geneviève,
June 12, 1809

The second volume of *Mithridates* by Adelung and Vater, a treatise on languages, has appeared in German. It deals with research connected to your theories.

ALEXANDER VON HUMBOLDT TO THE PRESIDENT OF THE AMERICAN PHILOSOPHICAL SOCIETY, PARIS, JUNE 12, 1809

Mr. President,

Permit me to send you my work on the barometric surveys of the Andean cordillera. I should be honored if the distinguished Society over which you preside would accept it as a humble token of my devotion and gratitude.

I have the honor to remain,
Sir, and greatly admired colleague,
Your very humble and obedient servant,
Baron von Humboldt
Paris, June 12, 1809

Please be so kind as to remember me in particular to Messrs. Wistar, Rush, Patterson, Thornton, Seybert, Peale, Woodhouse, Collin, Hare, John Vaughan, Mease, Ellicot and Barton-Smith, who showered me with kindness during my visit to Philadelphia.

ALEXANDER VON HUMBOLDT TO THOMAS JEFFERSON, PARIS, SEPTEMBER 23, 1810

Sir,

I have the honor to present you the fourth and fifth part of my work on Mexico as a humble expression of my profound veneration and respect. Although these works were written in circumstances somewhat unfavorable to the tranquility of my mind, I flatter myself that you may find in them the expression of those independent convictions which have inspired me all my life, and which I regard as a heritage that cannot be taken away from me. My thoughts often turn to Monticello, and I picture under the peaceful shade of a magnolia the statesman who has established the happiness of an entire world. Tears come to my eyes when I think that the most virtuous of men is also the happiest. For what can equal the happiness that you, Sir, must feel, surrounded by hard-working, enterprising citizens, worthy of the liberty which you have achieved and preserved for them.

I need say nothing about the worthy person who has brought you this gift of my tender devotion. Throughout his residence in Paris Mr. Warden has been a credit to his country for his honest and loyal conduct, his love for the sciences, and the esteem in which he has been held among men of good will. I do not know how General Armstrong alone could be mistaken about him. We hope that Mr. Warden will be returned to us.

Be assured, Sir, of my greatest respect. I repeat my request to receive as a gift from your own hands of your work on Virginia. I have owned it for fifteen years but I should like to be able to show my friends a copy with your dedication to me. This is what my vanity desires, and I make no apology for it.

Humboldt
Paris, September 23, 1810.

THOMAS JEFFERSON TO ALEXANDER VON HUMBOLDT, MONTICELLO, APRIL 14, 1811

My dear Baron
The interruption of our intercourse with France, for some time past, has prevented my writing to you. a conveyance now occurs, by mr Barlow or mr Warden, both of them going in a public capacity. it is the first safe opportunity offered of acknoleging your favor of Sep[tember] 23. and the reciept at different times of the III.d part of your valuable work, 2.d 3.d 4.th & 5.th livraisons, and the IV.th part, 2.d 3.d & 4.th livraisons, with the Tableaux de la nature, and an interesting map of New Spain. for these magnificent & much esteemed favors accept my sincere thanks. they give us a knolege of that country more accurate than I believe we possess of Europe, the seat of the science of a thousand years. it comes out too at a moment

when those countries are beginning to be interesting to the whole world. they are now becoming the scenes of political revolution, to take their stations as integral members, of the great family of nations. all are now in insurrection. in several the Independants are already triumphant, and they will undoubtedly be so in all. what kind of government will they establish? how much liberty can they bear without intoxication? are their chiefs sufficiently enlightened to form a wellguarded government, and their people to watch their chiefs? have they mind enough to place their domesticated Indians on a footing with the whites? all these questions you can answer better than any other. I imagine they will copy our outlines of confederation & elective government, abolish distinction of ranks, bow the neck to their priests, & persevere in intolerantism. their greatest difficulty will be in the construction of their Executive. I suspect that, regardless of the experiment of France, and of that of the US. in 1784. they will begin with a Directory, and when the unavoidable schisms in that kind of Executive shall drive them to something else, their great question will come on, whether to substitute an Executive, elective for years, for life, or an hereditary one. but unless instruction can be spread among them more rapidly than experience promises, despotism may come upon them before they are qualified to save the ground they will have gained. could Napoleon obtain, at the close of the present war the independance of all the West India islands, & their establishment in a separate confederacy, our quarter of the globe would exhibit an enrapturing prospect into futurity. you will live to see much of this. I very little. I shall follow, however, chearfully my fellow laborers, contented with having borne a part in beginning this beatific reformation. I fear, from some expressions in your letter, that your personal interests have not been duly protected, while you were devoting your time, talents & labor for the information of mankind. I should sincerely regret it, for the honor of the governing powers, as well as from affectionate attachment to yourself, & the sincerest wishes for your felicity, fortunes and fame.

In sending you a copy of my Notes on Virginia, I do but obey the desire you have expressed. they must appear chetif enough to the author of the great work on South America. but from the widow her mite was welcomed, & you will add to this indulgence the acceptance of sencere assurances of constant friendship & respect.

Th: Jefferson

ALEXANDER VON HUMBOLDT TO THOMAS JEFFERSON, PARIS, DECEMBER 20, 1811

Sir,

Yesterday I arrived from Vienna, where my brother is a Minister of the King of Prussia, and where I have spent a month visiting my parents. Upon my return I was delighted to find the interesting letter you kindly sent me, Sir, and that you should accompany it with a gift to which I accord the greatest price. The Notes on Virginia will be placed in the library which my brother and I have established: it is a claim to glory for me to have enjoyed the benevolence, dare I say friendship, of a man who has excited the admiration of this century by his virtue and moderation. For fear that the frigate is about to depart, as they have warned me, I can add only a few lines. I am taking the liberty of sending you the final part of my report on Astronomical Observations and the sixth and seventh parts of the Essay on New Spain, with the corresponding atlases. I have sent, and have had my booksellers send, the previous parts by several mail routes: perhaps they have already reached you; I therefore beg you to tell me very frankly which parts are still missing from your set. I hope that communications will improve before long. I have completed two thirds of my work; the historical part is now at the printer's. Mr. Arrowsmith in London has stolen my large map of Mexico: and Mr. Pike has rather ungraciously taken advantage of the copy of this map which was undoubtedly passed to him in Washington: and besides, he has also misspelt all the names. I am sorry to have to complain about a citizen of the United States who has otherwise shown such bravery. My name does not appear in his book, and a quick glance at Mr. Pike's map will show you from where it was drawn. My fortune has suffered, not so much through my travels but through political upheavals; to lose one's fortune is to lose but little. I find consolation in my work, in memories, and in the esteem of those who recognize the purity of my intentions. Like you, I take the liveliest interest in the great struggle of Spanish America. It is hardly surprising that it is a bloody conflict, when one thinks that everywhere Man bears the stamp of the imperfection of social institutions, and that the peoples of Europe have for three centuries sought security in mutual resentment and the hatred of the classes. I shall not leave Europe until I have completed my work; the newspapers have me already in Tibet.

I am considering various projects, but I still prefer to go into Asia. This letter is brought to you by my friend Mr. Correa de Serra, a member of the Royal Society of London, and correspondent of the Institut who intends to settle in Philadelphia. He is a man of lofty, just and forceful spirit and is one of the greatest botanists of the century, although he has published only very little. I take the liberty of recommending him to your care, and beg you to recommend him to your friends in Philadelphia.

Please accept, my kind and honored friend, the expression of my admiration and gratitude.

Humboldt

Paris, The Observatory, December 20, 1811

THOMAS JEFFERSON TO ALEXANDER VON HUMBOLDT, [MONTPELIER], DECEMBER 6, 1813

My dear friend and Baron.

I have to acknolege your two letters of Dec[ember] 20. & 26. 1811. by mr Correa, and am first to thank you for making me acquainted with that most excellent character. he was so kind as to visit me at Monticello, and I found him one of the most learned and amiable of men. it was a subject of deep regret to separate from so much worth in the moment of it's becoming known to us.

the livraison of your Astronomical observations and the 6.th and 7.th on the subject of New Spain, with the corresponding Atlasses are duly recieved, as had been the preceding Cahiers. for these treasures of a learning so interesting to us, accept my sincere thanks. I think it most fortunate that your travels in those countries were so timed as to make them known to the world in the moment they were about to become actors on it's stage. that they will throw off their European dependance I have no doubt; but in what kind of government their revolution will end is not so certain. history, I believe furnishes no example of a priest-ridden people maintaining a free civil government. this marks the lowest grade of ignorance, of which their civil as well as religious leaders will always avail themselves for their own purposes. the vicinity of New Spain to the US. and their consequent intercourse may furnish schools for the higher, and example for the lower classes of their citizens. and Mexico, where we learn from you that men of science are not wanting, may revolutionise itself under better auspices than the Southern provinces. these last, I fear, must end in military despotism. the different casts of their inhabitants, their mutual hatreds and jealousies, their profound ignorance & bigotry, will be plaid off by cunning leaders, and each be made the instrument of enslaving the others. but of all this you can best judge, for in truth we have little knolege of them, to be depended on, but through you.

but in whatever governments they end, they will be *American* governments, no longer to be involved in the never-ceasing broils of Europe. the European nations constitute a separate division of the globe; their localities make them part of a distinct system; they have a set of interest of their own in which it is our business never to engage ourselves. America has a hemisphere to itself: it must have it's separate system of interests, which must not be subordinated to those of Europe. the insulated state in which nature has placed the American continent should so far avail it that no spark of war kindled in the other quarters of the globe should be wafted across the wide oceans which separate us from them. and it will be so. in 50. years more the US. alone will contain 50. millions of inhabitants; and 50. years are soon gone over. the peace of 1763. is within that period. I was then 20. years old, and of course remember well all the transactions of the war preceding it. and you will live to see the epoch now equally ahead of us, and the numbers which will then be spread over the other parts of the American hemisphere, catching long before that the principles of our portion of it, and concurring with us in the maintenance of the same system.—you see how readily we run into ages beyond the grave, and even those of us to whom that grave is already opening it's quiet bosom. I am anticipating events of which you will be the bearer to me in the Elysian fields 50. years hence.

You know, my friend, the benevolent plan we were pursuing here for the happiness of the Aboriginal inhabitants in our vicinities. we spared nothing to keep them at peace with one another, to teach them agriculture and the rudiments of the most necessary arts, and to encourage industry by establishing among them separate property. in this way they would have been enabled to subsist and multiply on a moderate scale of landed possession; they would have mixed their blood with ours and been amalgamated and identified with us within no distant period of time. on the commencement of our present war, we pressed on them the observance of peace and neutrality. but the interested and unprincipled policy of England has defeated all our labors for the salvation of these unfortunate people. they have seduced the greater part of the tribes, within our neighborhood, to take up the hatchet against us, and the cruel massacres they have committed on the women and children of our frontiers taken by surprise, will oblige us now to pursue them to extermination, or drive them to new seats beyond our reach, already we have driven their patrons & seducers into Montreal, and the opening season will force them to their last refuge, the walls of Quebec, we have cut off all possibility of intercourse and of mutual aid, and may pursue at our leisure whatever plan we find necessary to secure ourselves against the future effects of their savage and ruthless warfare. the confirmed brutalisation, if not the extermination of this race in our America is therefore to form an additional chapter in the English history of the same colored man in Asia, and of the brethren of their own colour in Ireland and wherever else Anglo-mercantile cupidity can find a two-penny interest in deluging the earth with human blood.—but let us turn from the loathsome contemplation of degrading effects of commercial avarice.

That their Arrowsmith should have stolen your map of Mexico, was in the piratical spirit of his country. but I should be sincerely sorry if our Pike has made an ungenerous use of your candid communications here; and the more so as he died in the arms of victory gained over the enemies of his country. whatever he did was on a principle of enlarging knolege, and not for filthy shillings and pence of which he made none from that book. if what he has borrowed has any effect it will be to excite an appeal in his readers from his defective information to the copious volumes of it with which you have enriched the world. I am sorry he omitted even to acknolege the source of his information. it has been an oversight, and not at all in the spirit of his generous nature. let me sollicit your forgiveness then of a deceased hero, of an honest and zealous patriot, who lived and died for his country.

You will find it inconcievable that Lewis's journey to the Pacific should not yet have appeared; nor is it in my power to tell you the reason. the measures taken by his surviving companion Clarke, for the publication, have not answered our wishes in point of dispatch. I think however, from what I have heard, that the mere journal will be out within a few weeks in 2. vol.s 8.vo these I will take care to send you with the tobacco seed you desired, if it be possible for them to escape the thousand ships of our enemies spread over the ocean. the botanical & zoological discoveries of Lewis will probably experience greater delay, and become known to the world thro' other channels before that volume will be ready. the Atlas, I believe, waits on the leisure of the engraver.

Altho' I do not know whether you are now at Paris, or ranging the regions of Asia to acquire more knolege for the use of man, I cannot deny myself the gratification of an endeavor to recall myself to your recollection ~~and~~ of assuring you of my constant attachment, and of renewing to you the just tribute of my affectionate esteem & high respect and consideration.

Th: Jefferson

THOMAS JEFFERSON TO ALEXANDER VON HUMBOLDT, MONTICELLO, JUNE 13, 1817

Dear Sir

The reciept of your Distributio geographica plantarum, with the duty of thanking you for a work which sheds so much new and valuable light on botanical science, excites the desire also of presenting myself to your recollection, and of expressing to you those sentiments of high admiration and esteem, which, altho' long silent, have never slept. the physical information you have given us of a country hitherto so shamefully unknown, has come exactly in time to guide our understandings in the great political revolution now bringing it into prominence on the stage of the world. the issue of it's struggles, as they respect Spain, ~~are~~ is no longer matter of doubt. as it respects their own liberty, peace & happiness we cannot be quite so certain. whether the blinds of bigotry, the shackles of the priesthood, and the fascinating glare of rank and wealth give fair play to the common sense of the mass of their people, so far as to qualify them for self government, is what we do not know. perhaps our wishes may be stronger than our hopes. the first principle of republicanism is that the lex majoris partis is the fundamental law of every society of individuals of equal rights: to consider the will of the society enounced by the majority of a single vote as sacred as if unanimous, is the first of all lessons in importance, yet the last which is thoroughly learnt.

this law once disregarded, no other remains but that of force, which ends necessarily in military despotism. this has been the history of the French revolution; and I wish the understanding of our Southern brethren may be sufficiently enlarged and firm to see that their fate depends on it's sacred observance.

In our America, we are turning to public improvements. schools, roads and canals are every where either in operation or contemplation. the most gigantic undertaking yet proposed is that of New York for drawing the waters of Lake Erie into the Hudson. the distance is 353. miles, and the height to be surmounted 691 feet. the expence will be great; but it's effect incalculably powerful in favor of the Atlantic states. internal navigation by steam boats is rapidly spreading thro all our states, and that by sails and oars will ere long be looked back to as among the curiosities of antiquity. we count much too on it's efficacy in harbor defence; and it will soon be tried for navigation by sea. we consider this employment of the contributions which our citizens can spare, after feeding, and clothing, and lodging themselves comfortably, as more useful, more moral, and even more splendid, than that preferred by Europe, of destroying human life, labor and happiness.

I write this letter without knowing where it will find you. but wherever that may be, I am sure it will find you engaged in something instructive for man. if at Paris, you are of course in habits of society with mr Gallatin our worthy, our able and excellent minister, who will give you, from time to time, the details of the progress of a country in whose prosperity you are so good as to feel an interest, and in which your name is revered among those of the great worthies of the world. god bless you, and preserve you, long to enjoy the gratitude of your fellow men, and to be blessed with honors, health, and happiness.

Th: Jefferson

ALEXANDER VON HUMBOLDT TO THOMAS JEFFERSON, PARIS, FEBRUARY 22, 1825

Sir,

The affectionate kindness with which you have honored me during these long years prompts me to address these lines to you. Count Carlo Vidua, the son of a Minister of State whose wise administration is cherished by the Piedmontese, will be the bearer of my lasting and respectful admiration for you. This young traveler has already traveled through Europe as far as the Arctic Circle, the Crimea, Asia Minor, Greece, Palestine, and Upper Egypt; in the New World he is going to study intellectual progress and those free and learned institutions which you have helped to establish and spread from the Missouri to the furthest points of South America. I once had the good fortune to discuss with you, in the Presidential Mansion in Washington, the events that will change the face of the world, and that your wisdom had long anticipated. Count Vidua will rediscover in the Virginian citizen what I admired so greatly in the first magistrate of a great nation. I envy him that good fortune which he is worthy to enjoy.

I remain with the most respectful gratitude, Sir,
Your very humble and obedient servant,
Alexander Humboldt
Paris, February 22, 1825

Humboldt's Account of His American Travels, Written for the American Philosophical Society (1804)

HUMBOLDT'S AUTOBIOGRAPHICAL ACCOUNTS

On various occasions in his life, inspired by different motives and particular circumstances, Humboldt dedicated his time and effort to drafting autobiographical notes.[2] He wrote the oldest text, called "Notice sur la vie littéraire de Mr. de Humbold [*sic*], communiquée par lui même au Baron de Forell," in French during his stay in Spain in spring 1799 and submitted it, together with a memorial,[3] to King Carlos IV. The aim was to inform the Spanish authorities—from whom he expected to obtain the permission to visit their colonies in America—about his person as well as his objectives and plans related to his famous American project.[4]

During his first stay in Cuba from December 19, 1800, until March 15, 1801, he conceived of the idea to write his autobiography. Throughout his American expedition he returned to this project, and in Santa Fé de Bogotá he formulated some notes in German, four-and-a-half pages long, dated August 4, 1801, but unfortunately his narratives reached only until the year 1790.[5] In ulterior annotations, which Humboldt added to this text in November 1839, he noted the phrase "not to be published ever"—probably due to some critical comments about his friend and travel companion Georg Forster, as well as some reflections concerning himself. Other pages were found hidden in his travel diaries, which Humboldt titled "Zeitepochen meines Lebens" (Epochs of my life), and where, in the form of annotations without any elaboration, he summed up his life until February 1803, when they left Guayaquil for Acapulco.

The next document of this type is the narrative of his American expedition written in June 1804 at the end of his stay in the United States. This text, reproduced below, has great significance as the first and only summary of the entire expedition authored by Humboldt himself, and the freshness of the very recent memories he describes adds additional interest.

Several of his letters had a character similar to that of these autobiographical annotations, particularly those that were elaborated during his American expedition: they were written in order to make public the itinerary of the travel undertaken until that moment, as well as to disseminate the first scientific results he had obtained. These letters thus provided information regarding his famous voyage long before Humboldt started to publish his travel narrative—the *Relation historique* (*Personal Narrative*)—which constitutes the first part of his important work *Voyage aux regións équinoxiales du Nouveau Continent.*

A few weeks after his return to Europe, in October 1804, Humboldt started the lecture of a "relation abrégée" of his travels in the Institute of France in Paris. Motivated by this narrative, whose text has not been conserved, and with the help of several of his letters that had been printed in different journals, the French physician and natural scientist Jean Claude de Lamétherie, a friend of Humboldt and member of the Académie des Sciences, compiled a text which he published in the *Journal de Physique* under the title "Notice d'un voyage aux tropiques, executé par MM. Humboldt et Bonpland en 1799, 1800, 1801, 1802, 1803 et 1804."[6] As of yet this text has been translated only into German and Dutch. Based on the travel descriptions provided by Humboldt's letters that were included in several journals—though also not drafted by the explorer himself—is also the text "Alexander von Humboldts Reisen um die Welt und durch das Innere von Südamerika" (Alexander von Humboldt's travels around the world and through the interior of South Amerika), published in 1805 by the journalist Friedrich Wilhelm von Schütz and translated a few years later into Polish.[7]

Shortly afterward, a new occasion to write an autobiographical account arose: Humboldt's friend Marc Auguste Pictet, a physician and professor in Geneva, maintained close relations with England and offered his help to disseminate the Prussian's work in Great Britain. For this campaign, on January 3, 1806, Humboldt sent him a description of his life, written in French, with the title "Mes confessions"—probably alluding to the work by Jean-Jacques Rousseau—where he added the express desire that it should someday be read and disseminated.

At an advanced age, Humboldt once again began to formulate a summary of his life. This time his effort was motivated by an article about him published in the ninth edition of the *Brockhaus Encyclopedia*. This document summarizes in a concise and strictly chronological form the important dates in Humboldt's life and, above all, expresses the image of himself that he wanted to present in his last years.[8] Finally, at the end of his long life, Humboldt made some fragmentary annotations in chronological order titled "Chronologische Folge der Zeitepochen meines Lebens" (Chronological order of the epochs of my life).[9]

The autobiographical summary written in the United States and published below has some peculiarities that bear mention. First of all, it is notable that Humboldt concludes his narrative with his arrival in Europe, though at the time of writing he was still in America. This can be explained by the argument that the text was written under the impression of the finalization of his expedition and his expected immediate return to Europe, as well as by his desire to deliver and print the first description of the complete travels as soon as possible. Furthermore, probably due to his sense of urgency about drafting the text and his desire to make accessible certain information only in his later publications, he limited himself to a mere narrative of the expedition and did not include a description of his scientific research. It is also noteworthy that the article was written in the third person, a rather unusual practice for a personal travel narrative.

Another interesting aspect of this article is that Humboldt does not mention the topic of slavery, or any other controversial subject such as the exploitation of the Indian, or the negative impact of the missionary in Spanish America. It can be supposed that this has to do with the fact that the text originally was published for the North American public, and at the time of his visit, Humboldt was still rather cautious in referring to these issues, whereas in later years, especially when he became disenchanted with the development of politics in the United States, he expressed his critics directly and openly.

BARON HUMBOLDT.

THE following abstract of the American Travels of the celebrated baron Humboldt and his companion Bonpland, has been drawn up from notes which the former has kindly furnished, and will supersede the many very incorrect accounts hitherto published relative to this interesting object.[10] Baron Humboldt, having travelled from the year 1790, as a naturalist, through Germany, Poland, France, Switzerland, and through parts of England, Italy, Hungary, and Spain,[11] came to Paris in 1798, when he received an invitation, from the directors of the National Museum, to accompany captain Baudin in his voyage round the world. Citizen Alexander Aime Gourjon Bonpland, a native of La Rochelle, and brought up in the Paris Museum, was also to have accompanied them; when on the point of departing, the whole plan was suspended until a more favourable opportunity, owing to the recommencement of the war with Austria, and to the consequent want of funds.

Mr. Humboldt, who, from 1792, had conceived the plan of travelling through India at his own expence, with a view of adding to the knowledge of the sciences connected with Natural History, then resolved to follow the learned men, who had gone on the expedition to Egypt. His plan was to go to Algiers in the Swedish Frigate which carried the consul Skjöldebrand, to follow the caravan which goes from Algiers to Mecca, going through Egypt to Arabia, and thence by the Persian Gulph to the English East-India establishments.

The War which unexpectedly broke out in October, 1798, between France and the Barbary states,[12] and the troubles in the East, prevented Mr. Humboldt from embarking at Marseilles, where he had been fruitlessly two months waiting to proceed. Impatient at this delay, and continuing firm in his determination to go to Egypt, he went to Spain, hoping to pass more readily under the Spanish flag from Cartagena to Algiers and Tunis. He took with him the large collection of Philosophical, Chemical, and Astronomical instruments, which he had purchased in England and France.

From a happy concurrence of circumstances, he obtained in February 1799, from the Court of Madrid permission to visit the Spanish colonies of the Two Americas, a permission which was granted with a liberality and frankness, which was honourable to the Government and to a Philosophic Age. After a residence of some months at the Spanish court, during which time the King[13] showed a strong personal interest in the plan, Mr. Humboldt, in June 1799, left Europe, accompanied by Mr. Bonpland, who, to a profound knowledge in Botany and Zoology, added an indefatigable zeal. It is with this friend that Mr. Humboldt has accomplished, at his own expence, his travels in the two Hemispheres, by land and sea, probably the most expensive which any *individual* has ever undertaken.

These two travellers left La Coruña in the Spanish ship Pizarro, for the Canary Islands, where they ascended to the crater of the Peak of Teide, and made experiments on the analysis of the air. In July they arrived at the port of Cumana, in South America. In 1799, 1800, they visited the coast of Paria, the missions of the Chaymas Indians, the province of New Andalusia (a country which had been rent by the most dreadful earthquakes, the hottest, and yet the most healthy, in the world) of New Barcelona, of Venezuela, and of Spanish Guayana. In January 1800, they left Caracas to visit the beautiful valley of Aragua, where the great lake of Valencia recalls to the mind the views of the lake of Geneva, embellished by the majesty of the vegetation of the tropics. From Portocabello they crossed to the south, the immense plains of Calabozo, of Apure, and of the Orinoco, also Los Llanos, a desert similar to those of Africa, where in the shade (by the reverberation of heat) the thermometer of Reaumur rose to 35° and 37° (111 to 115 Fahrenheit). The level of the country for 2000 square leagues does not differ 5 inches. The sand every where represents the horizon of the sea, without vegetation: its dry bosom hides the crocodiles, and the torpid boa (a species of serpent). The travelling here (as in all Spanish America except Mexico) is performed on horseback. They passed whole days without seeing a palm-tree or the vestige of a human dwelling. At St. Fernando de Apure, in the provinces of Barinas, Mess. Humboldt and Bonpland began that fatiguing navigation of nearly 1000 marine leagues, executed in canoes, making a chart of the country by the assistance of chronometers, the satellites of Jupiter, and the lunar distances. They descended the river Apure, which empties itself into the Orinoco, in 7 degrees of latitude, They ascended the last river (passing the celebrated cataracts of Maipures and Atures) to the mouth of the Guaviare. From thence they ascended the small rivers of Atabapobapa, Tuamini and Temi. From the mission of Yavita they crossed by land to the sources of the famous Rio Negro, which La Condamine[14] saw, where it joins the Amazone, and which he calls a sea of fresh water. About 30 Indians carried the canoes through woods of Maní, Lecythis and Laurus Cinamomoides to the Caño (or creek) of Pemichin. It was by this small stream that the travellers entered the Rio Negro, or Black River, which they descended to St. Carlos, which has been erroneously supposed to be placed under the equator, or just at the frontiers of Great Pará, in the government of Brazil. A canal from Temi to Pimichín, which from the level nature of the ground is very practicable, would present a fine internal communication between the Para and the province of Caracas, a communication infinitely shorter than that of Casiquiare.[15] From the fortress of St. Carlos on the Rio Negro, Mr. Humboldt went north up that river and the Casiquiare to the Orinoco, and on this river to the volcano Duida or the mission of the Esmeralda, near the sources of the Orinoco: the Indians Guaicas (a race of men almost pigmies, very white and very warlike) render fruitless any attempts to reach the sources themselves. From the Esmeralda Mess. Humboldt and Bonpland went down the Orinoco, when the waters rose, towards its mouth at St. Thomas de la Guayana, or the Angostura. It was during this long navigation that they were in a continued state of suffering, from want of nourishment, and shelter from the night rains, from living in the woods, from the mosquitoes and an infinite variety of stinging insects, and from the impossibility of bathing, owing to the fierceness of the crocodile and the little Carib fish, and finally the miasmata of the burning climate. They returned to Cumaná by the plains of Cari and the mission of the Carib Indians, a race of men very different from any other, and probably, after the Patagonians, the tallest and most robust in the world. After remaining some months at New Barcelona and Cumaná, the travellers arrived at the Havana, after a tedious and dangerous navigation, the vessel being in the night on the point of striking upon the Vibora rocks. Mr. Humboldt remained three months in the island of Cuba, where he occupied himself in ascertaining the longitude of the Havana, and in constructing stoves on the sugar plantations, which have since been pretty generally adopted. They were on the point of setting off for Veracruz, meaning, by the way of Mexico and Acapulco, to go to the Philippine Islands, and from thence, if it was possible, by Bombay and Aleppo, to Constantinople, when some false reports relative to Baudin's voyage alarmed them, and made them change their plan. The gazettes held out the idea that this navigator would proceed from France to Buenos Aires, and from thence, by Cape Horn, for Chili and the coast of Peru. Mr. Humboldt had promised to Mr. Baudin and to the Museum of Paris, that wherever he might be, he would endeavour to join the Expedition, as soon as he should know of its having been commenced. He flattered himself that his researches, and those of his friend Bonpland, might be more useful to science, if united to the labours of the learned men who would accompany captain Baudin.

These considerations induced Mr. Humboldt to send his manuscripts, for 1799 and 1800 direct to Europe and to freight a small schooner at Batabano, intending to go to Cartagena, and from thence, as quickly as possible, by the Isthmus of Panama, to the South Sea. He hoped to find captain Baudin at Guayaquil, or at Lima, and with him to visit New Holland, and the islands of the Pacific Ocean, equally interesting in a moral point of view, as by the luxuriance of their vegetation. It appeared imprudent to expose the manuscripts and collections already made to the risks of this proposed navigation. These manuscripts, of the fate of which Mr. Humboldt remained ignorant during three years, and until his arrival in Philadelphia, arrived safe, but one third part of the collection was lost by shipwreck. Fortunately (except the insects of the Orinoco and of the Rio Negro) they were only duplicates; but unhappily friar Juan González, monk of the order of St. Francis, the friend to whom they were entrusted, perished with them. He was a young man full of ardour, who had penetrated into this unknown world of Spanish Guayana further than any other European. Mr. Humboldt left Batabano in March 1801, and passed to the south of the island of Cuba on which he determined many geographical positions. The passage was rendered very long by calms, and the currents carried the little schooner too much to the west, to the mouths of the Atracto. The vessel put into the river Sinu, where no botanist had ever before visited, and they had a very difficult passage up to Cartagena. The season being too far advanced for the South Sea navigation, the project of crossing the isthmus was abandoned; and animated by the desire of being acquainted with the celebrated Mutis,[16] and admiring his immensely rich collections of objects of Natural History, Mr. Humboldt determined to pass some weeks in the woods of Turbaco, and to ascend (which took forty days) the beautiful river of Magdalena, of the course of which he sketched a chart.

From Honda, our travellers ascended through forests of oaks, of *melastoma*, and of *cinchona* (the tree which affords the Peruvian bark), to Santa Fé de Bogotá, capital of the kingdom of New Grenada, situated in a fine plain, elevated 1360 toises (of six French feet) above the level of the sea. The superb collections of Mutis, the majestic cataract of the Tequendama (falls of 98 toises height), the mines of Mariquita, St. Ana, and of Zipaquirá, the natural bridge of Icononzo (three stones thrown together in the manner of an arch, by an earthquake), these curious objects stopped the progress of Mess. Humboldt and Bonpland until the month of September 1801.

At this time, notwithstanding the rainy season had commenced, they undertook the journey to Quito, and passed the Andes of Quindío, which are snowy mountains covered with palmiers à cire (wax palm-trees) passeflores (passion flower) of the growth of trees, storax, and bambusa (bamboo). They were, during 13 days, obliged to pass on foot through places dreadfully swampy, and without any traces of population.

From the village of Cartago, in the valley of Cauca, they followed the course of the Chocó, the country of platina, which was there found in round pieces of basalt and green rock (Greinstein of Werner), and fossil wood. They pass by Buga to Popayan, a bishop's see, and situated near the volcanoes of Sotara and Puracé, a most picturesque situation, and enjoying the most delicious climate in the world, the thermometer of Réaumur keeping constantly at 16° to 18°

(68 to 72 Fahrenheit). They ascended to the crater of the volcano of Puracé, whose mouth, in the middle of snow, throws out vapours of sulphureous hydrogene with continued and frightful rumbling.

From Popayan they passed by the dangerous defiles of Almaguer (avoiding the infected and contagious valley of Patia), to Pasto and from this town, even now situated at the foot of a burning volcano, by Tuqueres and the province of Pastos, a flat portion of country, fertile in European grain, but elevated move than 1500 to 1600 toises above the towns of Ibarra and Quito.

They arrived, in January 1802, at this beautiful capital, celebrated by the labours of the illustrious La Condamine, of Bouguer, Godin, Jorge Juan, and Ulloa, and still more celebrated by the great amiability of its inhabitants, and their happy turn for the arts.

They remained nearly a year in the kingdom of Quito; the height of its snow-capped mountains, its terrible earthquakes (that of February 7, 1797, swallowed up 42,000 inhabitants, in a few seconds), its fertility, and the manners of its inhabitants, combined to render it the most interesting spot in the universe. After three vain attempts, they twice succeeded in ascending to the crater of the volcano of Pichincha, taking with them electrometers, barometers, and hygrometers. La Condamine could only stop here a few minutes, and that without instruments. In his time, this immense crater was cold and filled with snow. Our travellers found it inflamed, distressing information for the town of Quito, which is distant from it only 5000 to 6000 toises.

They made separate visits to the snowy and porphyritic mountains of Antisana, Cotopaxi, Tunguragué, and Chimborazo, the last the highest point of our globe.[17] They studied the geological part of the Cordillera of the Andes, on which subject nothing has been published in Europe, mineralogy (if the expression may be used) having been created, as it were, since the time of La Condamine. The geodesical measurements proved that some mountains, particularly the volcano of Tunguragué, has considerably lowered since 1750, which result agrees with the observations made to them by the inhabitants. During the whole of this part of the journey, they were accompanied by Mr. Charles Montúfar, son of the marquis of Selva Alegre of Quito, a person zealous for the progress of science, and who is, at his own expence, rebuilding the pyramids of Yaruquí, the extremity of the celebrated bases of the triangles of the Spanish and French academicians.[18] This interesting young man having followed Mr. Humboldt in the remainder of his journey through Peru and the kingdom of New Spain, is now on his passage with him to Europe.

Circumstances were so favourable to the efforts of the three travellers, that at Antisana they ascended 2200 French feet, and at Chimborazo, on 22 June 1802, nearly 3200 feet higher than La Condamine was able to carry his instruments. They ascended to 3036 toises elevation above the level of the sea, the blood starting from their eyes, lips, and gums. An opening, of 80 toises deep, and very wide, prevented them from reaching the top, from which they were only distant 134 toises.

It was at Quito that Mr. Humboldt received a letter from the National Institute of France,[19] informing him, that captain Baudin had proceeded by the Cape of Good Hope, and that there was no longer any hope of joining him.

After having examined the country overturned by the earthquake of Riobamba in 1797, they passed by the Andes of Azuay to Cuenca. The desire of comparing the barks (cinchona) discovered by Mr. Mutis at Santa Fe de Bogota, and with those of Popayan, and the cuspa and cuspare of New Andalusia and of the river Caroni (named falsely Cortex Augosturae), with the cinchona (barks) of Loja and Peru, they preferred deviating from the beaten track from Cuenca to Lima; but they passed with immense difficulties in the carriage of their instruments and collections, by the forest (Páramo) of Saraguro to Loja, and from thence to the province of Jaén de Bracamoros. They had to cross 35 times in two days the river Huancabamba, so dangerous for its sudden freshes. They saw the ruins of the superb Inca road (comparable to the finest roads in France, and which went upon the ridge of the Andes from Cusco to the Azuay, accommodated with fountains and taverns).

They descended the river Chamaya, which led them into that of the Amazones, and they navigated this last river down to the cataracts of Tomependa, one of the most fertile, but one of the hottest, climates of the habitable globe. From the Amazone River they returned to the south-east by the Cordilleras of the Andes to Montán, where they found they had passed the magnetic equator, the inclination being 0, although at seven degrees of south latitude. They visited the mines of Hualgayoc, where native silver is found at the height of 2000 toises. Some of the veins of these mines contain petrified shells, and which, with those of Pasco and Huantajaya are actually the richest of Peru. From Cajamarca they descended to Trujillo, in the neighbourhood of which are found the ruins of the immense Peruvian city, Mansiche.

It was on this western descent of the Andes that the three voyagers, for the first time, had the pleasure of seeing the Pacific Ocean. They followed its barren sides, formerly watered by the canals of the Incas at Santa, Huarmey and Lima. They remained some months in this interesting capital of Peru, of which the inhabitants are distinguished by the vivacity of their genius, and the liberality of their ideas.

Mr. Humboldt had the good fortune to observe the end of the passage of Mercury over the sun's disk, in the port of Callao. He was astonished to find, at such a distance from Europe, the most recent productions in chemistry, mathematics, and medicine; and he found great activity of mind in the inhabitants, who, in a climate where it never either rains or thunders, have been falsely accused of indolence.

From Lima our travellers passed by sea to Guayaquil, situated on the brink of a river, where the growth, of the palm tree is beautiful beyond description. They every moment heard the rumbling of the volcano of Cotopaxi, which made an alarming explosion on the 6 January 1803. They immediately set off to visit it a second time, when the unexpected intelligence of the speedy departure of the frigate Atlante determined them to return, after being seven days exposed to the dreadful attacks of the mosquitoes of Babahoyo and Ujibar.

They had a fortunate passage, by the Pacific Ocean, to Acapulco, the western port of the kingdom of New Spain, famous for the beauty of its harbour, which appears to have been formed by earthquakes, for the misery of its inhabitants, and for its climate, which is equally hot and unhealthy.

Mr. Humboldt had originally the intention to remain only four months in Mexico, and to hasten his return to Europe; his voyage had already been too much protracted, his instruments, particularly the chronometers, began to be out of order, and every effort that he made to have new ones sent to him proved of no avail. Add to this consideration that the progress of science is so rapid in Europe, that in a journey that lasts four or five years, great risk is run of contemplating the different phenomena under aspects, which are no longer interesting at the moment of publishing the result of your labours. Mr. Humboldt hoped to be in France in August or September 1803, but the attractions of a country, so beautiful and so varied, as is that of the kingdom of New Spain, the great hospitality of its inhabitants, and the fear of the yellow fever, so fatal from June to November, for those who come from the mountainous parts of the country, led him to stay a year in this Kingdom . . .

Our travellers ascended from Acapulco to Taxco, celebrated for its mines, as interesting as they are ancient. They rise, by small degrees, from the ardent valleys of Mezcala and Papagayo, where the thermometer of Réaumur stands in the shade, constantly from 28° to 31° (95 to 101 Fahrenheit), in a region 6 or 700 toises above the level of the Bea, where you find the oaks, the pines, and the fuguere (fern) as large as trees, and where the European grains are cultivated. They passed by Taxco, by Cuernavaca to the capital of Mexico. This city of 150,000 inhabitants is placed upon the ancient site of Tenochtitlán, between the lakes of Texcoco and Xochimilco (lakes which have lessened somewhat since the Spaniards have opened the canal of Huehuetoca) in sight of two snow-topped mountains, of which one (Popocatépetl), is even now an active volcano, surrounded by a great number of walks, shaded with trees, and by Indian villages.

This capital of Mexico, situated 1160 toises *above the sea*, in a mild and temperate climate, may doubtless be compared to some of the finest towns in Europe. Great scientific establishments, such the Academy of Painting, Sculpture, and Engraving, the College of Mines, (owing to the liberality of the Company of Miners of Mexico), and the Botanic Garden, are institutions which do honour to the government which has created them.

After remaining some months in the valley of Mexico, and after fixing the longitude of the capital, which had been laid down with an error of nearly two degrees, our travellers visited the mines of Moran and Real del Monte, and the Cerro of Oyamel, where the ancient Mexicans had the manufactory of knives made of the obsidian stone. They soon after passed by Querétaro and Salamanca to Guanajuato, a town of 50,000 inhabitants, and celebrated for its mines, more rich than those of Potosi have ever been. The mine of the count of Valenciana, which is 1840 French feet perpendicular depth, is the deepest and richest mine of the universe. This mine alone gives to its proprietor nearly 600.000 dollars annual and constant profit.

From Guanajuato they returned by the valley of Santiago to Valladolid, in the ancient kingdom of Michoacán, one of the most fertile and charming provinces of the kingdom. They descended from Pátzcuaro towards the coast of the Pacific Ocean to the plains of Jorullo, where, in 1759, in one night, a volcano arose from the level, surrounded by two thousand small mouths, from whence smoke still continues to issue. They arrived almost to the bottom of the crater of the great volcano of Jorullo, of which they analized the air, and found it strongly impregnated with carbonic acid. They returned to Mexico by the valley of Toluca, and visited the volcano, to the highest point of which they ascended 14,400 French feet above the level of the sea.

In the months of January and February 1804, they pursued their researches on the eastern descent of the Cordilleras, they measured the mountains Nevados de la Puebla, Popocatépetl, Iztaccihuatli, the great peak of Orizaba, and the Cofre de Perote; upon the top of this last Mr. Humboldt observed the meridian height of the sun. In fine after some residence at Jalapa, they embarked at Veracruz, for the Havana. They resumed the collections they had left there in 1801, and by the way of Philadelphia, embarked for France, in July 1804, after six years of absence and labours.[20] A collection of 6000 different species of plants (of which a great part are new) and numerous mineralogical, astronomical, chemical, and moral observations, have been the result of this expedition. Mr. Humboldt gives the highest praises to the liberal protection granted to his researches by the Spanish Government.

Baron Humboldt was born in Prussia, on the 14 of September 1769.

Notes

ABBREVIATIONS

AVH — *Alexander von Humboldt und die Vereinigten Staaten von Amerika: Briefwechsel*, edited by Ingo Schwarz

FE — *The Writings of Thomas Jefferson*, edited by Paul Leicester Ford (Ford Edition)

ME — *The Writings of Thomas Jefferson*, edited by Andrew A. Lipscomb and Albert Ellery Bergh for the Thomas Jefferson Memorial Association (Memorial Edition)

PEKNS — Alexander von Humboldt, *Political Essay on the Kingdom of New Spain*

PN — Alexander von Humboldt and Aimé Bonpland, *Personal Narrative of Travels to the Equinoctial Regions of the New Continent, during the Years 1799–1804*

PTJ — *Papers of Thomas Jefferson*, Main Series

PTJ-D — *Papers of Thomas Jefferson*, Digital Edition

TJP — Thomas Jefferson Papers, Library of Congress

INTRODUCTION

1. English translation of this text: www.columbia.edu/acis/ets/CCREAD/etscc/kant.html.

2. Jefferson to John Hollins, February 19, 1809, TJP.

3. From the 1950s onward, several publications have treated different aspects of their relationship and the exchange of information they established (see Terra, "Motives and Consequences," 314–16; Terra, "Studies of Documentation" 136–41, 560–68; Terra, "Humboldt's Correspondence," 783–806; Wassermann, "Six Unpublished Letters," 191–200; Lange, "Aus dem Briefwechsel," 32–45; Schwarz, "From Humboldt's Correspondence," 1–20; Rebok, "Two Exponents of the Enlightenment," 126–52; Rebok, "Transatlantic Dialogue," 329–69; and Caspar, "Young Man from 'Ultima Thule,'" 247–62). Some of these studies focus particularly on Humboldt's travel itinerary in the United States, his motives to come to this country, his scientific interests and activities, as well as his view of the political system in this part of America (see Friis, "Humboldts Besuch," 142–95; Friis, "Baron Humboldt's Visit," 1–35;

Schoenwaldt, "Humboldt und die Vereinigten Staaten," 431–82; Schwarz, "Humboldt's Visit to Washington," 43–56; and Théodoridès, "Les séjour aux Etats-Unis," 287–304). Other studies center around Humboldt's general vision of the United States in comparison to colonial Spanish America (see Schwarz, "Humboldts Bild von Latein- und Angloamerika," 1142–54; Schwarz, "Humboldt—Socio-Political Views"; and Rebok, "New Approach," 61–88). In recent years, several works have discussed Humboldt's impact in the United States, the importance of the transatlantic collaboration through scientific networks he established, and his general influence in the fields of literature, philosophy, and history (see, for instance, Schwarz, *Humboldt und die Vereinigten Staaten;* Sachs, *Humboldt Current;* Dassow Walls, *Passage to Cosmos;* Clark and Lubrich, *Transatlantic Echoes;* and Clark and Lubrich, *Cosmos and Colonialism*).

While scholarship in this field often deals with specific aspects of Jefferson and Humboldt's personal encounter or their correspondence, their contact and epistolary exchange can be analyzed and interpreted in a far larger scope, such as the transatlantic dialogue, the history of ideas, the progress of science, or in the context of the study of the Enlightenment, the institution of slavery, revolutions, or the situation of the Native American population in the United States. Four comparative studies between other Germans and Jefferson have been published recently, including some cases in which the two have not met: Friedrich Wilhelm von Geismar, Johann Wolfgang von Goethe, Wilhelm von Humboldt, and Klemens von Metternich (see Krippendorff, *Jefferson und Goethe;* Nicolaisen, "Jefferson and Friedrich Wilhelm von Geismar," 1–27; Herbst, "Jefferson und Wilhelm von Humboldt," 273–87; and Sofka, *Metternich, Jefferson and the Enlightenment*).

4. Jefferson had formulated clear ideas on the future of his country long before he took his position in Paris (see Palmer, "Dubious Democrat," 388–404).

1. BIOGRAPHICAL BACKGROUNDS

1. Wilhelm von Humboldt (1767–1835) was a Prussian philosopher, linguist, government functionary, diplomat, and the founder of the University of Berlin. The German university system, with its concept of the unity of research and freedom of teaching and learning, inspired by W. von Humboldt, had a marked impact on many American universities created during the nineteenth century.

2. For biographical information about Humboldt in English, see Stoddard, *Life, Travels and Books;* Bruhns, *Life of Humboldt;* Terra, *Life and Times of Humboldt;* Botting, *Humboldt and the Cosmos;* Kellner, *Alexander von Humboldt;* and Beck, *Alexander von Humboldt.* Among the latest publications in English are Helferich, *Humboldt's Cosmos;* Rupke, *Alexander von Humboldt;* and Macrory, *Nature's Interpreter.* A good source for the latest research results is the *International Review for Humboldtian Studies, HiN: Alexander von Humboldt im Netz:* www.uni-potsdam.de/u/romanistik/humboldt/hin.

Regarding his American expedition; see also Kutzinski, Ette, and Dassow Walls, *Alexander von Humboldt and the Americas.*

3. Kant's publication, considered one of the most influential works in the history of philosophy, marks the beginning of modern philosophy. German idealism is noteworthy for its systematic treatment of all the major branches of philosophy, including logic, metaphysics and epistemology, moral and political philosophy, and aesthetics, as part of a general system of philosophy. Fundamental to German idealism is the belief that the properties we discover in objects depend on the way those objects appear to us as perceiving subjects, and not something they possess "in themselves," apart from our experience of them.

4. Humboldt, *Florae fribergensis specimen.*

5. The marine and French explorer Nicolas Baudin was well known for his explorations in the Indian Ocean, the Canary Islands, and the Antilles. In October 1800, he undertook an expedition to New Holland, Australia. From there he went to Tasmania and Timor, and on his way home he stopped at the Island of France (now the Republic of Mauritius), where he died of tuberculosis in September 1803.

6. Aimé Goujand Bonpland was a French botanist and medical doctor (see Bell, *Life in Shadow;* and Schneppen, *Aimé Bonpland;* Foucault, *Pêcheur d'orchidées*).

7. Most of these early works were published only in German: Humboldt, *Mineralogische Beobachtungen;* Humboldt, *Versuche über die gereizte Muskelund Nervenfaser;* Humboldt, *Über die unterirdischen Gasarten;* Humboldt, *Versuche über die chemische Zerlegung des Luftkreises.*

8. Regarding his stay in Tenerife, see Hernández González, *Alejandro de Humboldt.*

9. Carlos Montúfar y Larrea (1778–1816) was, from Quito onward, the third permanent member of Humboldt's expedition. After their arrival in Paris, he went to Madrid to serve in the Spanish army. Later he returned to Quito with the order to fight against agitators, but he instead formed an alliance with the rebels and appointed his father, the marquis of Selva Alegre, as president of the Junta Suprema de Gobierno, created in 1810, which declared independence from Spain one year later. Together with Simón Bolívar, he entered Bogotá in a triumphal parade in December 1814. In 1816, he was captured and shot in Buga in the Battle of Tambo (see Hampe Martínez, "Carlos Montúfar y Larrea," 711–20).

10. Lack, *Humboldt and the Botanical Exploration.*

11. Rebok, "New Approach," 61–88.

12. An excellent overview of the content of his opus as well as the numerous editions and translations can be found in Fiedler and Leitner, *Humboldts Schriften.*

13. Oppitz, "Name der Brüder Humboldt," 277–429; Helferich, *Humboldt's Cosmos,* 345–46.

14. Representation of nature.

15. Humboldt to Johann Georg von Cotta, October 31, 1854, in Leitner, *Humboldt und Cotta*, 545.

16. This description of his life is based on the brief biography of Jefferson on Monticello's website: www.monticello.org/site/jefferson/brief-biography-thomas-jefferson. Among the considerable number of Jefferson biographies, the following works can be particularly recommended: Bernstein, *Thomas Jefferson;* Appleby, *Thomas Jefferson;* Burstein, *Inner Jefferson;* Peterson, *Jefferson and the New Nation;* Cunningham, *In Pursuit of Reason;* and Malone, *Jefferson and His Time.*

17. Recommended editions of Jefferson's *Notes on the State of Virginia* have been published by Shuffelton (1999) and Peden (1982).

18. Gerbi, *Dispute of the New World.*

19. Concerning Jefferson's time in France, see W. Adams, *Paris Years of Jefferson;* Barlow Callen, "Jefferson and France"; Kaplan, *Jefferson and France;* Jefferson, *Jefferson's European Travel Diaries;* Palmer, "Dubious Democrat," 388–404; and Moore and Moore, *Jefferson's Journey.*

20. Jefferson to James Monroe, January 13, 1803, TJP.

21. Among the recommended literature regarding the Louisiana Purchase are Cerami, *Jefferson's Great Gamble;* Kukla, *Wilderness So Immense;* Kennedy, *Jefferson's Lost Cause;* Kastor, *Louisiana Purchase;* Lewis, *Louisiana Purchase;* and M. Adams, "Jefferson's Reaction," 173–88.

22. For more information on the Corps of Discovery, see Ronda, *Jefferson's West;* Cutright, *Lewis and Clark;* and Jackson, *Jefferson and the Stony Mountains.*

2. HUMBOLDT'S VISIT TO THE UNITED STATES

1. Vincent F. Gray to James Madison, April 28, 1804, in Friis, "Humboldts Besuch," 146.

2. Vincent F. Gray to James Madison, May 8, 1804, ibid.

3. See Stagg, *Borderlines in Borderlands.*

4. Humboldt to Jefferson, May 24, 1804. All the letters between Humboldt and Jefferson are reproduced in the appendix of this book.

5. See the custom declaration form in *AVH*, 483.

6. A very good, detailed, and well-documented description of Humboldt's travels through the United States can be found in Friis, "Humboldts Besuch." An English version of this article, limited to his stay in Washington, is Friis, "Baron Humboldt's Visit." These articles contain interesting comments and impressions by several persons who met Humboldt.

7. Jefferson was elected as member of the Society in 1780, appointed councillor in 1781 and again in 1783, vice president in 1791, and president in 1797, a position he held until he resigned in 1815.

8. *Gaceta de Madrid*, no. 61, July 31, 1804, 677.

9. A precise description of the encounter between Humboldt and Jefferson can be found in Charles Willson Peale's personal annotations: Peale, *Selected Papers*, 2:690–710 and 5:326–43. Regarding Peale's importance as a painter during his own time, see L. Miller, "Charles Willson Peale," 47–68.

10. Peale, *Selected Papers*, 5:332.

11. Ibid., 2:690.

12. Joseph Elgar Jr. to Jefferson, November 24, 1801, *PTJ-D*, 35:717.

13. At that time, it was still called the President's House; the first president to name it the White House was Teddy Roosevelt in 1901.

14. Peale, *Selected Papers*, 2:693.

15. Smith, *The First Forty Years of Washington Society*, 396–97.

16. Humboldt to James Madison, June 19 and 20, 1804 (French original: "Les jours que j'ai passé parmi Vous à Washington ont été des plus délicieux de ma vie. C'est une jouissance morale, au dessus de tout ce que offer la nature physique, de se voir, s'etendre et de simpathiser dans les sentimens sur le bonheur social").

17. Humboldt to Jefferson, September 23, 1810: "Je me transporte souvent dans ma pensée à Monticello, je crois voir, à l'ombre paisable d'un Magnolia, l'homme d'état qui a fondé le bonheur d'un monde entier."

18. See Friis, "Humboldts Besuch," 182–83.

19. Peale, Fothergill, Woodhouse, and Collin had already departed Washington on June 9 for Annapolis, before returning to Philadelphia.

20. See Humboldt to Albert Gallatin, June 20, 1804, *AVH*, 95; and Albert Gallatin to Humboldt, June 27, 1804, ibid., 99.

21. Humboldt to James Madison, June 19 and 20, 1804, ibid., 94.

22. See document 4, ibid., 496.

23. Humboldt to James Madison, June 19 and 20, 1804, ibid., 94 (French original: "Il me paraît que je reverrai ce beau pays en peu d'années. Le chemin du Missouri aux côté de l'Océan pacifique seara alors déjà ouvert. . . . Avec de la santé et du courage tout cela pourrait s'exécuter").

24. Humboldt to Albert Gallatin, June 29, 1804, ibid., 95 (French original: "Quelque regret que j'ai de quitter sit tôt ce beau pays, où les progrès de l'esprit humain et la liberté civile présentent un spectacle aussi brillant").

25. Humboldt to William Thornton, June 20, 1804, ibid., 96 (French original: "J'espère que nous nous y reverrons un jour. Ce pays qui s'étend à l'ouest des montagnes présente un vaste champs à conquérir pour les sciences").

26. Humboldt to John Vaughan, June 10, 1805, ibid., 105.

27. James Madison to Humboldt, March 12, 1833, ibid., 186.

28. William Thornton to John Vaughan, July 6, 1804, ibid., 193.

29. Benjamin Smith Barton to Jefferson, May 28, 1804, TJP. See also Ewan and Ewan, *Benjamin Smith Barton*, 444–49.

30. See Friis, "Humboldts Besuch," 158.

31. Ibid., 167.

32. Ibid., 181.

33. Albert Gallatin to Hannah Gallatin, June 6, 1804, cited ibid., 176.

34. Cited ibid., 175.

35. Mrs. Samuel Harrison Smith [Margaret Bayard Smith] to Mary Ann Smith, June 19, 1804, cited ibid., 179.

36. Smith, *The First Forty Years of Washington Society*, 395–96.

37. Humboldt to John Vaughan, June 30, 1804, *AVH*, 105 (French original:

On a déjà imprimé bien des choses sur mon Expédition qui no sont pas correctes et ceci fixera mieux les époques").

38. "Original Communication—Supplementary," *Literary Magazine and American Register* (1804): 2:321–27. This document is held at the American Philosophical Society in Philadelphia.

39. See the copy of the membership diploma in *AVH*, 513–14. His brother, Wilhelm von Humboldt, received this same distinction in 1822.

40. See the list in Terra, "Studies of Documentation," 141; and Schoenwaldt, *Humboldt und die Vereinigten Staaten*, 466–67.

41. For instance, in a letter sent on November 25, 1802, from Lima to a member of the National Institute in Paris, he mentions that he will be back in Europe in September or October 1803.

42. Margot Faak's edition of Humboldt's travel diaries has been published over many years by the Alexander von Humboldt Research Center of the Academy of Sciences in Berlin. The first volume, *Lateinamerika am Vorabend der Unabhängigkeitsrevolution* (1982), is an anthology of his views on various topics. In the next volumes, Faak published a chronological selection of Humboldt's travel account under the title *Reise auf dem Rio Magdalena, durch die Anden und durch Mexiko* (vols. 8 and 9, published in 1990 and 2003); she concludes her edition with the latest volume, *Reise durch Venezuela* (2000). In 2005, Ulrike Leitner published a once-missing part of his journal that she found in Cracow (Humboldt, *Von Mexiko-Stadt nach Veracruz*).

43. This refers to the publication of the more descriptive and narrative parts of his diaries (see Humboldt, *Reise auf dem Rio Magdalena*, 394–402). Other parts, not yet published, include additional notes, scientific results of his measurements, and data he copied from other sources. In *Reise auf dem Rio Magdalena*, we can find annotations regarding this country, which he might have taken out of the libraries and archives he consulted during his visit, but they could also have been added later. They reveal Humboldt's interest in various subjects related to the United States, such as news on the Lewis and Clark expedition and the Sioux Indians (IV, 33V), the population of North America and other statistics taken out of the *Morse's American Gazeteer* of Boston from 1797 (124V), the west coast from Vancouver to Alaska (VIII 91R), comparisons of the measurements taken by Mackenzie and Franklin (VIII, 198V–200R), and ancient drawings on rocks (II and VI 219V) (see *Humboldts amerikanische Reisejournale*).

44. Humboldt, *Reise auf dem Rio Magdalena*, 397–98 (French original: "Je me sentais très émû. Me voir périr á la veille de tant de jouissance, voir périr avec moi tous les fruits de mes travaux, être la cause de la mort des deux personnes qui m'accompagnaient, périr dans un voyage de Philadelphie qui ne paraissait pas de toute nécessité (quoique entrepris pour sauver nos manuscripts et collections contre la perfide politique espagnole").

45. It has not been possible to identify this place (see ibid., 330).

46. Quoted in Smith, *The First Forty Years of Washington Society*, 395.

3. TRANSATLANTIC EXPERIENCES

1. Jefferson to Edward Rutledge, August 6, 1787, *PTJ*, 11:701.

2. Humboldt to Jefferson, June 27, 1804 (French original: "Je parts parce que ma position l'exige, mais j'emporte avec moi la Consolation, que tandisque l'Europe présente un spectacle imorale et mélancholique, le peuple de ce Continent marche à grands pas vers la perfection de l'état social. Je me flatte que je jouirai un jour de nouveau de cet aspect consolant, je simpathise avec Vous dans l'espérance . . . que l'humanité peut s'attendre à une grande amélioration par le nouvel Ordre des choses qui règne ici").

3. *PN*, 3:65.

4. Humboldt, *Views of Nature*.

5. Humboldt, *Lateinamerika am Vorabend*, 63–64 (French original: "D'où vient ce manque de moralité, d'où viennent ces soufrances, ce malaise dans lequel tout homme sensible se trouve dans les Colonies européennes? C'est que l'idée de la Colonie même est une idée immorale, c'est l'idée d'un pays qu'on rend tributaire à une autre, d'un pays dans lequel on ne doit parvenir qu'à un certain degré de prospérité, dans lequel l'industrie, les lumière ne doivent se répandre que jusqu'à un certain point. . . . Tout Gouvernement Colonial est un gouvernement de méfiance. On y distribue l'autorité non selon que la félicité publique des habitants l'exige, mais selon le soupçon que cette autorité peut s'unir, s'attacher trop au bien de la Colonie, devenir dangereux aux intérêts de la mère patrie").

6. Ibid., 64 (French original: "Nullepart un Européen doit avoir plus honte de l'être que dans les Isles, soit Françaises, soit Anglaises, soit Danoises, soit Espagnol[e]s. Se disputer quelle Nation traite les Nègres avec plus d'humanité c´est se moquer du mot humanité et demander s'il est plus doux d'éventré ou écorché").

7. Ibid., 12s.

8. Humboldt to Jefferson, June 12, 1809 (French original: "Mon livre a été dédié au Roi Charles IV pour calmer par là l'humeur que le Gouvernement de Madrid auroit pu montrer contre quelques individus à Mexico qui m'ont fournit plus de renseignemens que peutêtre la Cour auroit voulu").

9. Alejandro Malaspina, 1754–1810, marine officer and explorer, undertook a scientific expedition to various parts of the world for the Spanish government. He was imprisoned from 1795 to 1803 in La Coruña and later exiled to Italy.

10. *PN*, 1:41.

11. Humboldt, *Reise durch Venezuela*, 58.

12. In spite of this self-censorship, his published writings (mostly his work on Cuba and his essay on New Spain) contain several critical comments on colonialism, slavery, oppression of the Indians, and many more topics.

13. For example, in one letter to Jefferson Davis, dated March 24, 1857, Humboldt writes: "I can only offer you on my part the frank and lively gratitude of an old man of 88 years who considers himself half an Amer-

ican" (*AVH*, 418). For more on this issue, see Biermann and Schwarz, "Humboldt—'Half an American,'" 43–50.

14. Humboldt to Klemens von Metternich, May 28, 1836, *AVH*, 196 (original version: "Presque Américan moi meme . . .").

15. Quoted in Foner, *Humboldt on Slavery*, 335.

16. For more information on Humboldt and slavery, see Foner, *Humboldt on Slavery*; and Schwarz, "Shelter for a Reasonable Freedom," 169–82; also, Michael Zeuske has published two good articles in Spanish: "Humboldt y la comparación de las esclavitudes" and "Humboldt, esclavitud, autonomismo."

17. The original French version titled *Essai politique sur l'ile de Cuba* was published in Paris in 1826; a recent English edition appeared in 2011: Humboldt, *Political Essay on the Island of Cuba*.

18. Humboldt, *Cosmos*, 1:358. All references used in this work refer to the *Cosmos* edition of 1858. A newer edition was published in 1997.

19. *PN*, 7:269.

20. The documents can be found in *AVH*, 560–62.

21. Humboldt, *Briefe von Alexander von Humboldt*, 332 (translated from German by Ingo Schwarz).

22. Berghaus, *Briefwechsel Humboldt's mit Heinrich Berghaus*, 1:16–17 (translated from German by Ingo Schwarz).

23. Humboldt, *Briefe von Alexander von Humboldt*, 305 (translated from German by Ingo Schwarz).

24. Schoenwaldt, *Humboldt und die Vereinigten Staaten*, n123.

25. Friedrich von Gerolt to Humboldt, May 17, 1858, *AVH*, 422.

26. Humboldt to John Matthews, October 12, 1858, ibid., 462.

27. Jefferson to Charles Bellini, September 30, 1785, *PTJ*, 8:568–69.

28. Jefferson to John Banister, October 15, 1785, in Jefferson, *Life and Selected Writings*, 359–60.

29. Jefferson to James Monroe, June 17, 1785, ibid., 341–42.

30. Jefferson to George Wythe, August 13, 1786, ibid., 366. Other letters that offer revealing comments on his views of Europe are Jefferson to Anne W. Bingham, February 7, 1787, ibid., 11:122–23; and Jefferson to George Washington, May 2, 1788, *PTJ*, 13:128.

31. Jefferson to Charles Bellini, September 30, 1785, *PTJ*, 8:568–69.

32. See, for example, Stanton, "Those Who Labor"; J. Miller, *Wolf by the Ears*; Stanton, *Free Some Day*; Gordon-Reed, *The Hemingses of Monticello*; Stanton, *Slavery at Monticello*; Finkelman, "Jefferson and Slavery"; Wills, *Negro President*; Onuf, "To Declare Them a Free"; and Jordan, *White over Black*. See also the statement on the much-debated topic of Thomas Jefferson and slavery by the International Center for Jefferson Studies: www.monticello.org/site/plantation-and-slavery/thomas-jefferson-and-slavery. This website also offers a very useful list of Jefferson's quotations on slavery and emancipation: www.monticello.org/site/jefferson/quotations-slavery-and-emancipation. An interesting approach to the understanding of Jefferson's allegedly paradoxical and sphinx-like character is found in Valsania, *Limits of Optimism*. See also Bailyn, "Jefferson and the Ambiguities," 498–515.

33. Argument in the case of Howell vs. Netherland, April 1770, in FE, 1:376.

34. Jefferson, *Notes on the State of Virginia*, ed. Peden, 163.

35. Jefferson to John Holmes, April 22, 1820, FE, 10:157. Ford erroneously transcribed "ear" as "ears."

36. *PTJ*, 1:130.

37. Jefferson, *Notes on the State of Virginia*, ed. Peden, 143.

38. Stanton, *Free Some Day*, 11.

39. Jefferson, *Life and Selected Writings*, 49.

40. Jefferson, *Jefferson's European Travel Diaries*, 42.

41. Jefferson to Thomas Cooper, September 10, 1814, ME, 14:183.

4. A TRANSATLANTIC NETWORK OF KNOWLEDGE AND IDEAS

1. Humboldt to Jefferson, May 24, 1804 (French original: "Arrivé depuis le Mexique sur le sol heureux de cette République dont le Pouvoir exécutif à été confié á Vos lumières, c'est un doux devoir pour moi de Vous présenter mes respects et l"hommage de la haute admirations que Vos écrits, Vos actions et la liberté de Vos idées m'ont inspiré dès ma plus tendre jeunesse. Malgré le désir ardent que j´ai de revoir Paris, où j'ai travaillé longtem[p]s avec les C. C. Vauquelin et Chaptal, et où nous comptons publier nos travaux (Fruits de cette Expédition) je n'ai pas pu résister à l'intérêt moral de voir les États-unis et de jouir de l'aspect consolant d'un peuple, qui sait apprécier le don précieux de la Liberté").

2. Humboldt to Zaccheus Collins, May 20, 1804, *AVH*, 87 (French original: "Étant déjà très circonscript dans mons tems et n'étant venu aux États unis que par l'intérêt moral de voir un pays aussi sagement gouverné, je désire infiniment de pouvoir passer le plûtot possible à Philadelphie)."

3. Humboldt to James Madison, May 24, 1804, ibid., 91 (French original: "Pénétré de l'intérêt le plus vif pour la prosperité de l"espèce humaine sur le Sol des États unis, fruit de Votre sage legislation et des vertus civique de Vos Magistrats. . . . C'est una idée bien consolante pour moi qu'après avoir été témoin des grands phénomènes qui présente la Nature magestueuse de la Cordillère des Andes, qu'après avoir vu ce qui est grand dans le monde physique, je puisee jouir du Spectacle moral qui présente un people libre et digne de sa belle destinée").

4. French original: "Le désir de me rendre utile aux sciences physique et d'étudier l'homme dans ses différents états de barbarie et de culture m'a fait entreprendre, à mes propres fraix, en 1799 une Expédition aux Tropiques."

5. It was important for him to make that clear from the beginning, thus he also repeated it in a letter he wrote to James Madison the same day (ibid., 91).

6. Named after Luigi Galvani, the term "galvanism" refers to the study of contraction of a muscle stimulated by an electric current.

7. Ebeling's important work on the geography and history of North America (*Erdbeschreibung und Geschichte von Amerika*) has received much attention.

8. Letter dated July 30, 1795, *PTJ-D*, 28:424.

9. The toise is an old unit of length that was defined from 1799 to 1812 as

equivalent to 1.949 meters. Thus, 1,700 toises correspond to 3,313.3 meters or 10,870.41 ft. (French original: "J'aimerais Vous parler encore d'un objet que Vous avez si ingénieusement traité dans Votre ouvrage sur la Virginie, des dents de Mammot que nous avons découvert dans les Andes de l'Hémisphère austral à 1700 toises de hauteur sur l'Océon Pacifique").

10. See also Humboldt's letter of departure to Jefferson from June 27, 1804, in which he writes, "I have had the good fortune to see the first Magistrate of this great republic living with the simplicity of a philosopher" (French original: "J'ai eu le bonheur de voir le premier Magistrat de cette grande République vivre avec la simplicité d'un Siècle philosophique)."

11. Biermann, "War Humboldt ein 'Freiherr'?"

12. Jefferson to Humboldt, May 28, 1804.

13. Jefferson to Miguel Lardizabel y Uribe, July 6, 1787, *PTJ-D*, 11:553–54.

14. Ponce interprets this as Jefferson's lack of knowledge of and interest in the Hispanic territories: ("Fragmentos de un discurso," 5–24).

15. Jefferson to Humboldt, June 9, 1804.

16. Chinard, "Jefferson and the American Philosophical Society," 263–76.

17. Humboldt, "Tablas geográfico-políticas," 635–57.

18. The entire document can be seen in the Library of Congress, and it is published in *AVH*, 484–94.

19. Jefferson to Wistar, June 7, 1804, TJP.

20. *PEKNS* (Mexican edition), xvi–xx.

21. An instructive study of this controversy is Labastida, "Humboldt, México y Estados Unidos," 131–47. Schwarz, "Shelter for a Reasonable Freedom," 176–82.

22. Gallatin, "Tabellarische Übersicht," 328–34.

23. Humboldt, "Fortschritte in der Kultur," 320–28.

24. Harris and Buckley, *Zebulon Pike*.

25. Humboldt to Jefferson, December 20, 1811 (French original: Mr. Arrowsmith à Londres m'a volé ma grande Carte du Mexique: Mr. Pike a profité d'une manière peu généreuse de la communication qui lui a été faite sans doute a Washington de la copie de ma Carte: d'ailleurs il a estropié tous les noms. Je suis affligé d'avoir à me plaindre d'un citoyen des États Unis qui d'ailleurs a déployé un si beau courage. Mon nom ne se trouve pas dans son livre et un léger coup d'oeil sur la Carte de Mr. Pike Vous prouve d'où il a puisé").

26. Jefferson to Humboldt, December 6, 1813.

27. To William Dunbar, he wrote: "While Captain Lewis' mission was preparing . . . I knew that a thousand accidents might happen . . . in such a journey as his, and thus deprive us of the principal object of the expedition, to wit, the ascertaining the geography of that river, [and] I set myself to consider whether in making observations at land, that would furnish what we want of it at sea obliges us to supply by the time-keeper. . . . Before his confirmation of the idea however, Captain Lewis was gone. In conversation afterwards with Baron Humboldt, he observed that the idea was correct, but not new; and that I would find it in the third volume of Delalande" (Jefferson to William Dunbar, May 25, 1805, ME, 11:77).

28. Sellers, *Mr. Peale's Museum.*

29. Peale, *Selected Papers,* 2:694.

30. Jefferson to Isaac Briggs, June 11, 1804, TJP.

31. Humboldt to Jefferson, February 22, 1825 (French original: "J'ai eu le bonheur de m'entretenir avec Vous dans le Palais de la Présidence à Washington sur des événemens qui vont changer la face du monde et que Votre sagacité avait devinés depuis longtemps").

32. Humboldt to Jefferson, June 27, 1804 (French original: "J'ai eu le bonheur de voir le premier Magistrat de cette grande République vivre avec la simplicité d'un Siècle philosophique et me revoir avec cette bienfaisante bonté, qui attache pour jamais").

33. See www.monticello.org/site/research-and-collections/number-letters-jefferson-wrote.

34. In fact, we know from the references in their correspondence that Humboldt had sent one lost letter from Bremen, mentioned in his next letter dated May 3, 1808, and another one on December 26, 1811, to which Jefferson refers in his letter of December 6, 1813. See also Jefferson's epistolary record "Summary Journal of Letters," maintained from November 11, 1783, to June 25, 1826.

35. Humboldt to Jefferson, May 30, 1808 (French original: "Au milieu des malheurs auxquels a succombé ma patrie j'ai tâché de Vous exprimer de tems en tems les sentiments de reconnaissance et d'admiration dont je suis pénétré pour Vous").

36. Humboldt to Jefferson, June 12, 1809 (French original: "Je n'ai pas été heureux depuis que j'ai quitté Votre beau pays").

37. Ibid. (French original: "Quelle carrière que la Votre! Quel exemple ravissant, Vous avez donné, d'énergie de caractère, de douceur et de profondeur dans les affections les plus tendres de l'âme, de modération et de justice comme premier magistrat d'un Etat puissant! Ce qui a été crée par Vous, Vous le voyez prospérer. Votre retraite à Monticello est un événement, dont la mémoire ne s'éteindra jamais dans les fastes de l'humanité").

38. Humboldt to Jefferson, December 20, 1811 (French original: "j'ose dire de l'amitié d'un homme qui a etonné ce siècle par ses vertus et sa modération").

39. Jefferson to Humboldt, June 13, 1817.

40. Jefferson to Humboldt, April 14, 1811.

41. Humboldt to Jefferson, December 20, 1811 (French original: "Je suis vivement intéressé comme Vous à la grande lutte de l'Amérique espagnole. Il faut pas s'étonner que la lutte soit sanglante, lorsqu'on pense que les hommes portent par tout l'empreinte de l'imperfection des institutions sociales et que les peuples d'Europe depuis trois siècles ont cherché leurs sécurité dans le ressentiments mutuel et la haine des Castes").

42. Jefferson to Humboldt, December 6, 1813.

43. Jefferson to Humboldt, June 13, 1817.

44. James R. Sofka, "'A Commerce Which Must Be Protected': The International Policy of Thomas Jefferson, 1785–1809," forthcoming, manuscript

provided by the author. See also Sofka, *Metternich, Jefferson, and the Enlightenment*, 262.

45. Sofka, "A Commerce Which Must Be Protected."

46. Jefferson to John Adams, May 17, 1818, TJP.

47. Jefferson to Humboldt, June 13, 1817.

48. Ponce, "Fragmentos de un discurso," 19.

49. Jefferson to Humboldt, March 6, 1809.

50. Jefferson to Humboldt, December 6, 1813.

51. The complex topic of Jefferson's attitude toward Indians is analyzed in chapter 8.

52. Humboldt to Jefferson, June 12, 1809 (French original: "Je félicite l'Etat du choix qu'on fait les citoyens de l'Amérique. Il m'a laissé une impression très belle. J'aime Votre expression 'il nous promet—a wise and honest administration.' Ce mot d'honnête renferme tout ce qui este juste, libéral, vertueux").

53. See particularly his *Personal Narrative*, but also his work on Mexico. An extensive composition of his comments on Indians can be found in his travel diaries edited by Faak and published in 1982.

54. Jefferson to Humboldt, December 6, 1813.

55. See, for instance, Humboldt to Jefferson, February 22, 1825, where he recommended to his protection the Italian traveler and author Carlo Conte de Vidua (1785–1830).

56. Humboldt to Jefferson, May 30, 1808.

57. *PEKNS*, 1:12–13.

58. For more information about David Bailie Warden and his relationship with Humboldt, see *AVH*, 22–23.

59. David Bailie Warden to Jefferson, July 24, 1808, in Sowerby, *Catalogue of the Library*, 4:291.

60. Jefferson to David Bailie Warden, February 25, 1809, ibid.

61. Humboldt to William Thornton, June 20, 1804, *AVH*, 96 (French original: "j'ai vu que c'est le seul coin de la terre où l'homme jouit de Sa liberté, et où de petits maux sont compensés par de grands bien").

62. Humboldt to William Thornton, June 20, 1804, ibid., 97 (French original: "Cette abominable loi qui permet l'importation des Nègres dans la Caroline méridionale est un opprobre pour un État, dans lequel je sais qu'ils existent des têtes très bien organisés. En suivant la seule marche qui dicte l'humanité, on exportera sans doute au commencement moins de Coton. Mais hélas! que je déteste cette Politique qui mesure et évalue la félicité publique simplement d'après la valeur des Exportations! Il est la richesse des Nations comme celle des Individus. Elle n'est que l'accessoire de notre félicité. Avant d'être libre, il faut être juste, et sans justice il n'y a pas prospérité durable").

63. Other North Americans expressed their disagreements with these theories, such as Benjamin Franklin during his time as American representative in Paris (1776–1882), Alexander Hamilton, and Thomas Paine in his works *Common Sense* (1776) and *Rights of Men* (1792), but not in such a detailed and scientific manner as Jefferson (see Gerbi, *Dispute of the New World*, 240–52).

64. *PEKNS*, 4:217.

65. Ibid., 1:vixxxi.

66. Ibid., 3:48.

67. Ibid, 3:37.

68. Humboldt, *Researches*, 1:4. See also the latest English edition of this work: Humboldt, *Views of the Cordilleras*.

69. Humboldt to Albert Gallatin, June 20, 1804, 298 (French original: "je préfère Votre Climat à tout autre. Car l'air le plus sain est celui où l'on respire le plus librement").

70. See, for instance, the letter to his brother Wilhelm von Humboldt, October 17, 1800, in Humboldt, *Briefe aus Amerika*, 105.

71. Humboldt to Antonio José Cavanilles, April 22, 1803, in Humboldt, *Briefe aus Amerika*, 225 (Original version in Spanish: "Han exagerado muchos Europeos la influencia de estos climas en el espíritu, y afirmando que es imposible soportar aquí el trabajo intelectual pero nosotros debemos publicar lo contrario, y decir por experiencia propia que nunca nos hemos hallado con más fuerzas que al contemplar las bellezas y magnificencia con que se presenta aquí la naturaleza. Su grandeza, sus infinitas y nuevas producciones nos electrizaban, por decirlo así, nos llenaban de alegría, y hacen invulnerables"). Regarding Humboldt's studies on climate, see also Humboldt, *Asie Centrale*.

72. For detailed information, see the chapter "De Pauw's First American Opponent," in Gerbi, *Dispute of the New World*, 194–288.

73. See Kohut, "Clavijero y las disputas," 52–81.

74. Gerbi, *Dispute of the New World*, 212.

75. For more information on Molina, see Orrego González, "Juan Ignacio Molina," 961–76.

76. Manuel de Salas, "Representación sobre el estado de la agricultura, industria y comercio del reino de Chile" (1796), cited in Gerbi, *Dispute of the New World*, 293.

77. For more information on Unánue, see Casalino, "Hipólito Unanue," 431–38.

78. Nieto Olarte, *Orden natural y orden social*.

79. Caldas, "Influjo del clima," 264–71. Regarding this article, see also Nieto Olarte, Castaño, and Ojeda, "Influjo del clima," 91–114.

80. Nieto Olarte, *Orden natural y orden social*, 197.

81. Years later, though, in his personal copy of the *Notes*, Jefferson annoted: "Cavigero [*sic*] says, "I do not remember that any American nation has any tradition of elephants or hippopotami, or other quadrupeds of equal size. I do not know if any of the numerous excavations made in New Spain has brought to light the carcass of a hippopotamus, or even the tooth of an elephant." For an English translation of Jefferson's quotation from the Italian text, see Jefferson, *Notes on the State of Virginia*, ed. Shuffelton, 302n57.

82. On this point, Jefferson was mistaken: The Novohispanian Clavijero was born in Mexico, and though he later lived and died in Bologna, Italy, he was not Italian.

83. Jefferson to Joseph Willard, March 24, 1789, *PTJ-D*, 14:697.

84. Jefferson to William Short, April 27, 1790, in *PTJ*, 16:388.

85. Jefferson to Charles Willson Peale, January 15, 1809, in Peale, *Selected Papers*, 2:1168–69.

86. For detailed information about Humboldt's works in Jefferson's library, see *AVH*, 21.

87. Jefferson to Humboldt, April 14, 1811.

88. A French translation published under the title *Observations sur la Virginie* received considerably less attention than the original English version (see Barker, "Unraveling the Strange History," 134–77).

89. Humboldt to Jefferson, June 12, 1809.

90. Humboldt to Jefferson, September 23, 1810.

91. He received the book on December 19, 1811.

92. See Jefferson's own copy of the *Notes on the State of Virginia* (London: Stockdale, 1787), Special Collections, University of Virginia, personal annotations on page 18: "Baron Humboldt states that in Lat. 37° (which is nearly over medium parallel) perpetual snow is no where known so low as 1200 toises = 7671 feet above the level of the sea, in sesquialtoral ratio nearly to the highest peak of Otter."

93. Ibid., see annotations on page 132: "a plant newly discovered by the great naturalist and traveller Baron Humboldt on the mountains of S. America, at the height of 2450 toises above the sea. . . . [T]he same scientific traveller, by analysis of the air, at different heights of the mountain of Chimborazo, which he ascended to the height of 3036 toises (546 toises higher than had ever been done by man before, and within 224 toises of its top) found that the oxygen being specifically heavier than the azotic part of the Atmosphere, it's proportion lessened in that ascent 27. or 28. to 19 ½ hundredths parts."

94. *PEKNS*, 2:231.

95. Ibid., 1:xiviii.

96. *PN*, 3:65, 68, 151.

97. Humboldt, *Researches*, 1:59.

98. "Chétif" translates from the French as "weak," "frail," "miserable." Jefferson to Humboldt, April 14, 1811.

99. Madame de Tessé to Jefferson, October 9, 1809, *PTJ-D*, 1:593 (French original: "Mr de la Fayette m'accusera de ceder a mon Gout seulement, Lorsque je crois Remplir un devoir d'équité en vous priant de placer dans votre bibliotheque La gravure d'un illustre voyageur, passionné de votre Gouvernement, et grand admirateur de votre personne. je ne doute point que Mr humbold ne soit tres flatté de se trouver á monticello quand il en aura connoissance, mais jai pourtant quoiqu'en puisse dire mon neveu moins d'envie de lui plaire par cet envoi, que de plaisir a le Recompenser").

100. Madame de Tessé to Jefferson, March 24, 1810, *PTJ-D*, 2:310.

101. Count Théodore Pahlen to Jefferson, June 25, 1810, *PTJ-D*, 2:487 (French original: "Monsieur, Si je n'avais été obligé d'attendre une occasion favorable pour faire parvenir en même tems la petite caisse ci-jointe qui contient un portrait de Mr le Baron de Humboldt et que Madame de Tessé

m'a particulièrement recommandé, connaissant l'intérêt que Vous prenez, Monsieur, à ce Savant voyageur").

102. John Vaughan to Jefferson, August 19, 1809, *PTJ-D*, 1:452.

103. Ibid., 1:455.

104. Jefferson to John Vaughan, Aug 31, 1809, *PTJ-D*, 1:482.

105. Marquis de Lafayette to Jefferson, March 12, 1811, *PTJ-D*, 3:444. He refers to the large Asian expedition Humboldt was planning in order to be able to compare his findings from the American and the Asian continents.

106. David Bailie Warden to Jefferson, December 10, 1811, *PTJ-D*, 4:325.

107. Cited in Terra, "Humboldt's Correspondence," 787.

108. Humboldt to William Thornton, June 10, 1804, *AVH*, 96 (French original: "Ce sont des grand phénomènes moraux qui ne perdent pas en les analisant de près et qui laissent une impression bienfaisante dans l'âme").

109. Humboldt to Alexis de Tocqueville, March 24, 1858 (Archive Tocqueville, Chateau de Tocqueville), cited ibid., 120.

110. See the original texts of Ralph Waldo Emerson, Henry David Thoreau, Edgar Allen Poe, Frederic Edwin Church, and Louis Agassiz, with references to Humboldt published in Clark and Lubrich, *Transatlantic Echoes;* and Clark and Lubrich, *Cosmos and Colonialism.*

111. Regarding Humboldt's legacy in American culture, literature, and art, see Dassow Walls, "Hero of Knowledge," 133.

112. See Dassow Walls, "Rediscovering Humboldt's Environmental Revolution" and *Seeing New Worlds.*

113. Mathewson, "Humboldt's Image and Influence," 416–38; Bunkse, "Humboldt and an Aestetic Tradition," 127–46.

114. Gould, "Church, Humboldt and Darwin," 94–107

115. Fiedler and Leitner, *Humboldts Schriften*, 397.

116. Among these visitors were John Lloyd Stephans, Bayard Taylor, Alexander Dallas Bache, George Folsom, Francis Lieber, Benjamin Silliman, Charles Patrick Daly, Moses Wight, Edward J. Young, George Catlin, George Ticknor, Daniel Huntington, and Maria Mitchell. For more information on these encounters as well as their personal impressions of Humboldt, see Schoenwaldt, *Humboldt und die Vereinigten Staaten*, 432–46.

117. See Thompson et al., "Proceedings: Humboldt Commemoration," 225–46. This article describes the immediate reaction to Humboldt's death in the United States; it consists of various letters containing personal memories of Humboldt that were read at a meeting of the American Geographical and Statistical Society in New York City on June 2, 1859.

118. For information on how the centennial of Humboldt's birth was celebrated in different places in the United States, see Nollendorfs, "Humboldt Centennial Celebration" 59–66; and Dassow Walls, *Passage to Cosmos*, 304–5.

119. A list of all the places in the world named after Humboldt is presented in Oppitz, "Name der Brüder Humboldt," 277–429.

120. See Baron and Seeger, "Moritz Hartmann in Kansas," 9.

121. John B. Floyd to Humboldt, July 14, 1858, *AVH*, 457 (French original:

"Nous ne saurions oublier vos services, ni les bienfaits que le monde a reçus de vous—Non seulement le nom de Humboldt est dans toutes les bouches, sur notre immense continent, des bords de l'Atlantique à ceux de la mer Pacifique mais en outre, nous en avons honoré nos rivières et plusieurs points de notre territoire et la posterité le retrouvera partout à côté des noms de Washington, Jefferson et Franklin").

122. See Nichols, "Why Was Humboldt Forgotten?," 399–415.

123. Dassow Walls, "Hero of Knowledge," 128.

124. For more information on the role of the descendants of German immigrants in keeping alive the memory of Humboldt in the United States, see Nichols, "Why Was Humboldt Forgotten?," 408–12; and Dassow Walls, *Passage to Cosmos*, 319–20.

125. See Clagett, *Scientific Jefferson Revealed*, appendix, 118.

5. JEFFERSON PRESENTS HIS NEW NATION

1. See Cogliano, *Jefferson: Reputation and Legacy*; and Wilson, "Behold Me at Length," 155–78.

2. Watts, "Jefferson, the "Encyclopédie,"" 318–25.

3. Among the European women Jefferson corresponded with were Maria Cosway, Adrienne Catherine de Noailles (Madame de Tessé), Baronne Anne Louise Germaine de Staël-Holstein, Elisabeth Françoise Sophie de Lalive de Bellegarde (Comtesse de Houdetot), Madame de Bréhan, and Anne Mangeot Ethis de Corny. For more information about his correspondence with particular women, see Chinard, *Trois amitiés françaises*, 9; Chinard, "Correspondence de Madame de Staël, 621–40; and Kaminski, *Jefferson in Love*.

4. Jefferson to John Randolph, August 25, 1775, *PTJ*, 1:241.

5. Jefferson to Marquis de Lafayette, March 10 and 12, 1781, ibid., 5:113, 129–30.

6. Jefferson to Marquis de Chastellux, January 16, 1784, ibid., 6:467.

7. Jefferson to Marquis de Chastellux, December 24, 1784, ibid., 7:580–82 and September 2, 1785, ibid., 8:467–70.

8. Jefferson to Marquis de Chastellux, June 7, 1785, ibid., 8:184–86.

9. Jefferson to Maria Cosway, October 12, 1786, ibid., 10:443–55.

10. Ibid., 10:447.

11. Ibid., 10:448.

12. Jefferson to Maria Cosway, May 21, 1789, ibid., 15:142–43.

13. Jefferson to Maria Cosway, June 23, 1790, ibid., 16:550–51.

14. Jefferson to Maria Cosway, December 24, 1786, ibid., 16:627; see also Jefferson to Maria Cosway, September 26, 1788, ibid., 13:638–39.

15. Jefferson to Madame de Bréhan, May 9, 1788, ibid., 13:150.

16. Jefferson to Madame de Tessé, September 6, 1795, ibid., vol. 28, 452.

17. A recent study on Jefferson's relationship with women is Kukla, *Mr. Jefferson's Women*.

18. Jefferson to Comte de Volney, December 9, 1795, *PTJ*, 28:551.

19. See Jefferson to Marquis de Lafayette, June 16, 1792, ibid., 24:85–86; and Jefferson to Thomas Paine, Philadelphia, June 19, 1792, ibid.

20. Jefferson to Philip Mazzei, April 24, 1796, ibid., 29:81–83.

21. Nevertheless, previous to these occurrences, in his letters to some people Jefferson was already cautious about the censorship they were going to face. See, for instance, Jefferson to Thomas Paine, December 23, 1788, ibid., 14:373.

22. For more biographical information about Tadeusz Kósciuszko, see Guthorn, "Kósciuszko as Military Cartographer," 49–53.

23. Jefferson to Tadeusz Kósciuszko, February 21, 1799, *PTJ*, 31:53.

24. Jefferson to Philip Mazzei, April 29, 1800, ibid., 31:544.

25. See, for instance, several letters sent to the French diplomat Edmond-Charles Genêt in 1793.

26. Jefferson to Tadeusz Kósciuszko, February 21, 1799, *PTJ*, 31:53.

27. He deemed, for instance, "Oriental" languages to be unimportant.

28. Jefferson to Pierre Samuel du Pont de Nemours, April 12, 1800, *PTJ*, 31:496.

29. Jefferson to Thomas Paine, March 18, 1801, FE, 8:18.

30. Jefferson to Joseph Priestley, June 19, 1802, ibid., 8:158.

31. See Smith, *The First Forty Years of Washington Society*, 395–97.

32. Jefferson to Pierre Samuel du Pont de Nemours, June 28, 1809, *PTJ-D*, 1:315.

33. Jefferson to Tadeusz Kósciuszko, February 26, 1810, ibid., 2:258.

34. Ibid., 2:257.

35. Jefferson to Tadeusz Kósciuszko, April 16, 1811, *PTJ-D*, 3:565.

36. Jefferson to Pierre Samuel du Pont de Nemours, April 15, 1811, ibid., 3:560.

37. See letters to Tadeusz Kósciuszko, June 28, 1812, FE, 9:361–64; to the Marquis de Lafayette, November 30, 1813, ibid., 9:434; and to Philipp Mazzei, December 29, 1813, ibid., 9:440–43.

38. Jefferson to Tadeusz Kósciuszko, June 28, 1812, ibid., 362.

39. Jefferson to the Marquis de Lafayette, November 30, 1813, ibid., 9:362.

40. Jefferson to Humboldt, June 13, 1817.

41. *PN*, 6:116.

42. Jefferson to Pierre Samuel du Pont de Nemours, April 24, 1816, FE, 10:22–26.

43. Jefferson to Marquis de Lafayette, December 26, 1820, ibid., 10:179–81.

44. Jefferson to Marquis de Lafayette, November 4, 1823, ibid., 10:279–83.

45. Chaconas, "Jefferson-Korais Correspondence," 65.

46. This term was coined by the French sociologist Pierre Bourdieu and refers to the resources that are available to an individual on the basis of the honor, prestige, or recognition one holds within a culture.

47. Humboldt to Albert Gallatin, June 27, 1804, *AVH*, 100.

6. TWO VIEWS OF THE HAITIAN REVOLUTION

1. About this topic, see, for instance, Hoffmann, Gewecke, and Fleischmann, *Haiti 1804—Lumières et ténèbres;* Yacou, *Saint-Domingue espagnol;* Geggus, "Naming of Haiti," 43–68; Geggus, *Impact of the Haitian Revolution;*

Geggus, *Haitian Revolutionary Studies;* and Hunt, *Haiti's Influence on Antebellum America.*

2. The position of the other Founding Fathers of the United States toward Haiti is the subject of the following articles: Brown, *Toussaint's Clause;* Sidbury, "Saint Domingue in Virginia," 531–52; Matthewson, "George Washington's Policy," 321–36; Matthewson, "John Adams and the Independence of Haiti."

3. Jefferson to Tench Coxe, June 1, 1795, TJP.

4. The Declaration of Independence: www.ushistory.org/declaration/document/index.htm.

5. Humboldt, *Lateinamerika am Vorabend,* 64. (French original: "Les Gouvernements européens ont si bien réussi à répandre la haine et la désunion dans les Colonies qu'on n'y connaît presque pas les plaisirs de la société du moins tout divertissement durable dans lequel beaucoup de familles doivent se réunir est impossible. De cette position naît une confusion d'idées et de sentiments inconcevables, une tendance révolutionnaire générale. Mais ce désir se borne à chasser les Européens et à se faire après la guerre entre eux").

6. Leitner, "Anciennes folies neptuniennes!"

7. See the explanations regarding these documents in Zeuske, "Humboldt y la comparación de las esclavitudes," 75–76.

8. *PEKNS,* 1:12.

9. Humboldt, *Island of Cuba* (1856), 186.

10. Ibid., 187.

11. Ibid., 396.

12. Ibid., 397.

13. Zeuske, "Humboldt y la comparación de las esclavitudes," 68.

14. Argote-Freyre, "Humboldt and Arango y Parreño," 273–80.

15. The following publications offer detailed information about the political situation and the different interests of the United States concerning France and Saint-Domingue, as well as Jefferson's attitude during these years: Matthewson, *Proslavery Foreign Policy;* Hickey, "America's Response," 361–79; Matthewson, "Jefferson and the Nonrecognition of Haiti," 22–48; Matthewson, "Jefferson and Haiti," 209–48; Auguste, "Jefferson et Haiti," 333–48; Wills, *Negro President,* 33–46; Brown, *Toussaint's Clause,* 179–99; Rebok, "La Revolution de Haïti," 75–95.

16. Jefferson to James Madison, February 5, 1799, *PTJ,* 30:9–11.

17. Regarding the conversation Jefferson had with the French representative in Washington, Louis André Pichon, about this subject in 1801, see Brown, *Toussaint's Clause,* 184ff.

18. In this context, it is interesting to contrast the opposed positions of Jefferson and Timothy Pickering, who reproached Jefferson for applying a double standard in the case of the French and the Haitian Revolutions (Wills, *Negro President,* 33–46). See also Hickey, "Timothy Pickering and the Haitian Slave Revolt," 149–63.

19. See Lokke, "Jefferson and the Leclerc Expedition," 322–28.

20. This letter is published in Auguste, "Jefferson et Haiti," 335–36.

21. See also Dubois, "Haitian Revolution and the Sale of Louisiana," 93–116.

22. Dubois, *Avengers of the New World*, 225.

23. Auguste, "Jefferson et Haiti," 346.

24. Egan, "United States, France, and West Florida," 227–52.

25. Auguste, "Jefferson et Haiti," 347.

26. For more information, see D. Egerton, *Gabriel's Rebellion;* and Nicholls, *Whispers of Rebellion.*

27. Jefferson to Rufus King, July 13, 1802, TJP.

28. Jefferson to James Monroe, November 24, 1801, *PTJ-D*, 35:720.

29. J. Miller, *Wolf by the Ears*, 126–29.

30. Humboldt to William Thornton, June 20, 1804, *AVH*, 96–97 (French original: "Plus que les événements récents de S[an] Domingue ont offusqué la vérité et plus il paraît du devoir de tout homme moral de replacer le problème dans son vrai jour").

7. ENGAGEMENT WITH THE NATURAL WORLD

1. Modern taxonomy introduced the rank of family between order and genus.

2. A detailed analysis of these different views and the evolution of their concepts over the years can be found in Sloan, "Buffon-Linnaeus Controversy," 356–75.

3. F. Egerton, "History of the Ecological Sciences, Part 1," 93.

4. Shugart and Woodward, *Global Change.*

5. Worster, *Nature's Economy.*

6. Fränzle, "Humboldt's Holistic World View," 57–90.

7. *Ideen zu einer Geographie der Pflanzen* (Ideas for a geography of plants) is based on a French text that Humboldt published in the same year under the title *Essai sur la géographie des plantes accompagné d'un tableau physique*, but because it contains numerous corrections and new additions, it cannot be seen as a mere translation.

8. Stearn, *Humboldt, Bonpland, Kunth*, 10. See also Fiedler and Leitner, *Humboldts Schriften*, 250–53.

9. Humboldt, *Essay on the Geography of Plants*, ed. Jackson, 79.

10. *PN*, 1:iii.

11. Humboldt, *Views of Nature*, 375.

12. Nicolson, "Humboldtian Plant Geography," 290.

13. Humboldt, *Cosmos*, 1:vii–viii.

14. Ibid., 60.

15. For more information on this issue, see Dettelbach, "Humboldt between Enlightenment and Romanticism," 9–20; Köchy, "Das Ganze der Natur"; Monreal and Álvarez Falcón, "Del racionalismo ilustrado," 349–57.

16. Jahn, "Humboldt's Cosmical View."

17. Rebok, "El arte al servicio de la ciencia."

18. Dassow Walls, "Rediscovering Humboldt's Environmental Revolution," 758–60; Dassow Walls, *Passage to Cosmos*, 8–9; Guha, *Environmentalism*, 26–27; Sachs, *Humboldt Current*.

19. *PEKNS*, 1:59–60.

20. Ibid., 1:22.

21. *PN*, 4:63–64, 142.

22. Dassow Walls, *Seeing New Worlds*, 134–47 ("Thoreau as Humboldtian"). See also Dassow Walls, "The Search for Humboldt," 473–77; and Schneider, *Thoreau's Sense of Place*.

23. Sachs, "Utimate Other," 114; Sachs, *Humboldt Current*, 338–53; see also Dassow Walls, "The Search for Humboldt," 473–77; and Sachs, "Humboldt's Legacy," 35.

24. F. Egerton, "History of the Ecological Sciences, Part 32," 253.

25. Jefferson to Pierre Samuel du Pont de Nemours, March 2, 1809, TJP.

26. For more information on Jefferson and his contribution to science, see Clagett, *Scientific Jefferson Revealed*; Thomson, *Passion for Nature*; Thomson, *Jefferson's Shadow*; Martin, *Jefferson: Scientist*; Bedini, *Jefferson, Statesman of Science*; Coonen and Porter, "Jefferson and American Biology," 745–50; West, "Jefferson as Scientist," 298–99; and Bedini, *Jefferson and Science*.

27. Jefferson, *Notes on the State of Virginia*, ed. Peden, 24.

28. Ibid., 25.

29. Thomson, *Passion for Nature*, 17.

30. Sowerby, *Catalogue of the Library*, 1:297–545. See also the long lists of books and pamphlets on agriculture, gardening, and botany in Jefferson's library, included in the appendix in Jefferson, *Thomas Jefferson's Garden Book*.

31. C. Miller, *Jefferson and Nature*, 1–4.

32. Jefferson to Adams, October 14, 1816, TJP.

33. C. Miller, *Jefferson and Nature*, 9.

34. Jefferson, *Jefferson's Memorandum Books*.

35. *PTJ-D*, 16:351.

36. For more information on Jefferson and weather observations, see www.monticello.org/site/research-and-collections/weather-observations.

37. Bedini, *Jefferson and Science*, 29–33.

38. Jefferson, *Notes on the State of Virginia*, ed. Shuffelton, 87.

39. For more information on Noah Webster, see Kendall, *Forgotten Founding Father*; and Mergen, *Snow in America*.

40. Jefferson to Wilson C. Nicholas, April 19, 1816, TJP.

41. Jefferson to Thomas Cooper, October 7, 1814, TJP.

42. Bedini, *Jefferson and Science*, 85. See also Smith Barton's article, in which he affirms that the information Jefferson contributed to this field is "equalled by that of few persons in the United States" (*Transactions*, 3:334–47). Besides the twinleaf *Jeffersonia diphylla*, a fungus and a mineral were also named after Jefferson.

43. FE, 7:477.

44. Hailman, *Jefferson on Wine*.

45. Hatch, *Gardens of Monticello*, 5. See also Hatch, *Fruits and Fruit Trees of Monticello*; and Hatch, "*A Rich Spot of Earth*."

46. Jefferson to Charles Willson Peale, August 20, 1811, *PTJ-D*, 4:93.

47. Jefferson, *Thomas Jefferson's Garden Book*, v–x.

48. See, for instance, letters written on October 31, 1803, October 26, 1805, and February 21, 1807. See also the list of plants that Jefferson sent from Paris about 1786 to Francis Eppes in the appendix of Jefferson, *Thomas Jefferson's Garden Book*.

49. Morgan, *Jefferson and the Natural World*, xiii. Regarding the beginning of the horticultural exchange between North America and France, see Wulf, *Brother Gardeners*.

50. Jefferson to Leonard Case, April 8, 1826, in Jefferson, *Thomas Jefferson's Garden Book*, 620.

51. Jefferson to Harry Innes, March 7, 1791, *PTJ-D*, 19:521.

52. Jefferson to Archibald Cary, January 7, 1787, *PTJ*, 9:158.

53. See Dugatkin, *Jefferson and the Giant Moose*; Lane, "Enlightened Controversy," 37–40; Wilson, "Jefferson, Buffon."

54. Jefferson to Buffon, October 1, 1787, *PTJ*, 12:194–95.

55. Bedini, *Jefferson and American Vertebrate Paleontology*.

56. Thomson, *Legacy of the Mastodon*.

57. Jefferson, *Notes on the State of Virginia*, ed. Peden, 55.

58. Boyd, "Megalonyx, the Megatherium," 420–35.

59. Jefferson to Archibald Stuart, May 26, 1796, *PTJ*, 29:113.

60. Jefferson, "Memoir on the Discovery," 246–60.

61. Wistar, "Description of the Bones," 526–31.

62. López Piñero and Glick, *El Megaterio de Bru*.

63. Cohen, *Science and Founding Fathers*, 292.

64. Rice, "Jefferson's Gift of Fossils," 597–627.

65. Robinson, "American Cabinet of Curiosities, 41–58.

66. More information on this topic can be found in: Wallace, "Jefferson and the Indians"; Johansen, "Franklin, Jefferson and American Indians"; Kennedy, "Jefferson and the Indians"; Bragaw, "Thomas Jefferson and the American Indian Nations"; and Sheehan, *Seeds of Extinction*.

67. Jefferson to Marquis de Castellux, June 7, 1785, *PTJ*, 8:185–86.

68. Jefferson, *Notes on the State of Virginia*, ed. Peden, 62.

69. See Thomson, *Passion for Nature*, 102.

70. Thomas Jefferson to Meriwether Lewis, June 20, 1803, Instructions, TJP.

71. Wallace, "Jefferson and the Indians," viii.

72. Humboldt, *Researches*, 1:53–60.

73. Ibid., 59.

74. FE, 8:iii.

75. Ingo Schwarz affirms the idea that Jefferson's *Notes on the State of Virginia* served as a model for Humboldt's regional descriptions (see Schwarz, "From Humboldt's Correspondence," 7).

76. Jefferson to Humboldt, April 14, 1811.

8. PARALLELS AND DISCREPANCIES

1. Kish, *Source Book in Geography*, 364.

2. Pattison, "Four Traditions of Geography," 202–6.

3. ME, 8:i–vii; Allen, "Imagining the West."

4. See Rebok, "Influence of Bernhard Varenius," 271–88.

5. ME, 13:iii.

6. It must be mentioned, however, that since not all of the Jefferson Papers have yet been published, the possibility exists that Jefferson did comment on Varenius's work. Presently there are two edition projects in progress: The Presidential Papers Project (1801–1809) at Princeton and the Retirement Paper Project (1809–1826) at the International Center for Jefferson Studies at Monticello, Charlottesville.

7. Sowerby, *Catalogue of the Library*, 85–356.

8. Humboldt, *Cosmos*, 1:67.

9. Ibid., 1:66.

10. A more detailed analysis of this question can be found in Rebok, "Alejandro de Humboldt y el modelo."

11. Sowerby, *Catalogue of the Library*, 254

12. While this idea cannot be explored in the frame of this work, it should provide inspiration for future studies.

13. Kish, *Source Book in Geography*, 370

14. About the predecessors and sources of Varenius, see Capel Saez, *Varenio, Geografía general*, 38–42.

15. Puig-Samper and Rebok, "Charles Darwin and Humboldt."

16. Jefferson to Archibald Stuart, December 23, 1791, *PTJ*, 22:436.

17. See McDonald, *Jefferson and the Power of Knowledge*.

18. See, for instance, Addis, *Jefferson's Vision for Education*, 68–87; and Cohen, *Science and Founding Fathers*.

19. Drouin, "Humboldt et la popularization," 45–63.

20. Numerous studies have been published on Jefferson and religion; see, for example, Conkin, "Religious Pilgrimage," 19–49; Sheridan, *Jefferson and Religion*; Addis, *Jefferson's Vision for Education*, esp. 68–87; Jefferson, *Jefferson's Extracts from the Gospels*; and Ragosta, *Religious Freedom*.

21. Jefferson to Humboldt, December 6, 1813.

22. See *AVH*, 61–62, for detailed information with examples from different letters.

23. Nollendorfs, "Humboldt Centennial Celebrations," 65.

24. Jefferson, "Description of a Mould-Board," 313–22. See also Stanton, "Better Tools," 200–222.

25. Shapley, "Notes on Thomas Jefferson," 234–37; Bedini, *Jefferson and Science*, 71–83.

26. Jefferson to Benjamin Waterhouse, March 3, 1818, FE, 10:103.

27. The entire speech is reproduced in *AVH*, 580.

28. Jefferson to Humboldt, March 6, 1809.

29. See the comments already cited in his December 6, 1813, letter to Humboldt.

30. Manning and Cogliano, *Atlantic Enlightenment*.

31. Himmelfarb, *Roads to Modernity*; Dunn, *Sister Revolutions*.

32. Bailyn, "Political Experience," 339.

33. Paret, "Jefferson and the Birth of European Liberalism," 491.

APPENDIX

1. Translation by Jeremy Rogers.

2. A larger version of this introduction has been published in Spanish, together with the Spanish translation of Humboldt's travel description: Puig-Samper and Rebok, "Humboldt y el relato," 69–84.

3. This document is published in Puig-Samper, "Humboldt, un prusiano," 329–55.

4. It is thus interesting to compare this document of introduction with the first letter Humboldt sent to Jefferson in order to present himself.

5. Published in Biermann and Lange, "Cómo Humboldt llegó a ser naturalista," 108–13.

6. 59 Year 12(/13) 2 (1804): 122–39.

7. Fiedler and Leitner, *Humboldts Schriften*, 28–33.

8. Published in *Die Gegenwart*, 749–62; as well as *Deutsche Lehr- und Wanderjahre*, 260–89.

9. His autobiographical notes were published in Humboldt, *Aus meinem Leben*.

10. In the following text, Humboldt's orthographical errors regarding the names of the places and persons have been retained.

11. This reference to Humboldt's possible stay in Spain cannot be documented and appears to be an error.

12. The Barbary States were Northwest African Berber states, many of which practiced state-supported piracy in order to obtain tribute from weaker Atlantic powers: the independent kingdoms Morocco, Algiers, Tunis, and Tripoli owed a loose allegiance to the Ottoman Empire. The United States fought two separate wars with them, the First Barbary War with Tripoli (1801–5) and the Second Barbary Way with Algiers (1815–16). The wars brought an end to the American practice of paying money to the pirate states and helped to end piracy in that region.

13. King Carlos IV received Humboldt in March 1799 in Aranjuez, after some arrangements made by the ambassador of Saxony in Madrid, Philippe von Forell, with Minister of State Mariano Luis de Urquijo, who became the principal supporter of Humboldt in the Spanish Court.

14. Charles-Marie de La Condamine (1701–1774) was a French mathematician and traveler who directed the geodesic expedition to Ecuador (1736–1743), in which participated the Spanish midshipmen Jorge Juan and Antonio de Ulloa.

15. The Casiquiare communicates with the fluvial system of the Amazon

and the Orinoco, as Humboldt was able to demonstrate officially, though the indigenous people were already aware of this connection. At the mid-nineteenth century, this area had been explored by the Frontier Commission directed by José de Iturriaga, in the course of which numerous settlements were founded that Humboldt later visited.

16. José Celestino Mutis y Bosio (1732–1808) was a medical doctor from Cadiz, professor of mathematics in the Colegio del Rosario in Santa Fé de Bogotá in 1762, and a correspondent of Linnaeus, who introduced the Newtonian physics in New Granada. From 1781 onward, he directed the Royal Botanical Expedition to New Granada and had eminent disciples such as Francisco José de Caldas, his own nephew Sinforoso Mutis, Jorge Tadeo Lozano, Francisco Antonio Zea, etc., many of them leaders in the independence movement in New Granada.

17. Humboldt thought that the Chimborazo was the highest mountain until he received the first information from Colonel Crawford about the highest of the Himalayan Mountains in 1807.

18. Between 1741 and 1743, the Spanish-French geodesical expedition erected some pyramids in the plains of Yaruquí under the direction of Charles-Marie de La Condamine that determined the base for his measurements of the meridian. The destruction of the pyramids was ordered by the court of Quito, since the collaboration of the Spanish scientists was not sufficiently highlighted.

19. Jean Baptiste Joseph Delambre to Humboldt, January 22, 1801.

20. Humboldt left France on January 4, 1799, and returned to Bordeaux on August 3, 1804.

Bibliography

Adams, Mary. "Jefferson's Reaction to the Treaty of San Ildefonso." *Journal of Southern History* 21 (1955): 173–88.

Adams, William Howard. *The Paris Years of Thomas Jefferson*. New Haven and London: Yale University Press, 1997.

Addis, Cameron. *Jefferson's Vision for Education, 1760–1845*. New York: Peter Lang, 2003.

Allen, John Logan. "Imagining the West: The View from Monticello." In *Thomas Jefferson and the Changing West: From Conquest to Conservation*, edited by James, P. Ronda, 3–23. Albuquerque: University of New Mexico Press, 1997.

Appleby, Joyce. *Thomas Jefferson*. New York: Times Books, 2003.

Argote-Freyre, Frank. "Humboldt and Arango y Parreño: A Dialogue." In Alexander von Humboldt, *Political Essay on the Island of Cuba*. Princeton: Markus Weiner; Kingston: Ian Randle, 2001.

Auguste, Yves. "Jefferson et Haiti." *Revue d'Histoire Diplomatique* (1973): 333–48.

Bailyn, Bernard. "Jefferson and the Ambiguities of Freedom." *Proceedings of the American Philosophical Society* 137, no. 4 (1993): 498–515.

———. "Political Experience and Enlightenment Ideas in Eighteenth-Century America." *American Historical Review* 67 (1961–62): 339–51.

Barker, Gordon S. "Unraveling the Strange History of Jefferson's *Observations sur la Virginie*." *Virginia Magazine of History and Biography* 112, no. 2 (2004): 134–77.

Barlow Callen, Mary Elisabeth. "Thomas Jefferson and France, 1784–89: Can Virtue Exist in a Luxurious World?" Master's thesis, University of Virginia, 1983.

Baron, Frank, and Scott Seeger. "Moritz Hartmann (1817–1900) in Kansas: A Forgotten German Pioneer of Lawrence and Humboldt." *Yearbook for German-American Studies* 39 (2004): 1–22.

Beck, Hanno. *Alexander von Humboldt*. Wiesbaden: Steiner, 1959–61.

Bedini, Silvio A. *Jefferson and Science*. Monticello Monograph Series. Charlottesville, Va.: Thomas Jefferson Foundation; Chapel Hill: University of North Carolina Press, 2002.

———. *Thomas Jefferson and American Vertebrate Paleontology*. Charlottesville: Commonwealth of Virginia, Department of Mines, Minerals and Energy, 1985.
———. *Thomas Jefferson, Statesman of Science*. New York: Macmillan, 1990.
Bell, Stephen. *A Life in Shadow: Aimé Bonpland in Southern South America, 1817–1858*. Stanford, Calif.: Stanford University Press, 2010.
Berghaus, Heinrich, ed. *Briefwechsel Alexander von Humboldt's mit Heinrich Berghaus aus den Jahren 1825 bis 1858*. 2 vols. Jena: Hermann Costenoble, 1869.
Bernstein, Richard B. *Thomas Jefferson*. New York: Oxford University Press, 2003.
Biermann, Kurt-Reinhard. "War Alexander von Humboldt ein "Freiherr" (oder "Baron")?" *Humboldt im Netz* 12, no. 23 (2011). www.unipotsdam.de/u/romanistik/humboldt/hin/hin23/biermann.htm.
Biermann, Kurt-Reinhard, and Fritz Lange. "Cómo Alejandro de Humboldt llegó a ser naturalista y explorador." In *Alejandro de Humboldt: Modelo en la lucha por el progreso y la liberación de la humanidad*, 108–13. Berlin: Akademie Verlag, 1969.
Biermann, Kurt-Reinhard, and Ingo Schwarz. "Alexander von Humboldt—'Half an American.'" *Alexander-von-Humboldt-Magazin* 67 (1996): 43–50.
Botting, Douglas. *Humboldt and the Cosmos*. New York: Harper and Row, 1973.
Boyd, Julian P. "The Megalonyx, the Megatherium, and Thomas Jefferson's Lapse of Memory." *Proceedings of the American Philosophical Society* 102, no. 5 (1958): 420–45.
Bragaw, Stephen G. "Thomas Jefferson and the American Indian Nations: Native American Sovereignty and the Marshall Court." *Journal of Supreme Court History* 31, no. 2 (2006): 155–80.
Brown, Gordon S. *Toussaint's Clause: The Founding Fathers and the Haitian Revolution*. Jackson: University Press of Mississippi, 2005.
Bruhns, Karl, comp. *Life of Alexander von Humboldt*. London: Longmans, Green, 1873.
Bunkse, Edmund V. "Humboldt and an Aesthetic Tradition in Geography." *Geographical Review* 71, no. 2 (1981): 127–46.
Burstein, Andrew. *The Inner Jefferson: Portrait of a Grieving Optimist*. Charlottesville and London: University Press of Virginia, 1996.
Caldas, José Francisco. "El influjo del clima sobre los seres organizados." *Semanario del Nuevo Reino de Granada*, no. 22 (1808): 200–207; no. 30 (1808): 264–71.
Cannon, Susan Faye. *Science in Culture: The Early Victorian Period*. New York: Dawson, 1978.
Capel Saez, Horacio, ed. *Bernhard Varenio, Geografía general, en la que se explican las propiedades generales de la tierra*. Barcelona: Ediciones de la Universidad, 1974.
Casalino, Carlota. "Hipólito Unanue: El poder político, la ciencia ilustrada

y la salud ambiental." *Revista Peruana de Medicina Experimental y Salud Publica* 25, no. 4 (2008): 431–38.
Caspar, Gerhard. "A Young Man from 'Ultima Thule' Visits Jefferson: Alexander von Humboldt in Philadelphia and Washington." *Proceedings of the American Philosophical Society* 155, no. 3 (2011): 247–62.
Cerami, Charles A. *Jefferson's Great Gamble: The Remarkable Story of Jefferson, Napoleon and the Men behind the Louisiana Purchase.* Naperville, Ill.: Sourcebooks, 2003.
Chaconas, Stephen G. "The Jefferson-Korais Correspondence." *Journal of Modern History* 14, no. 1 (1942): 64–70.
Chinard, Gilbert. "Jefferson and the American Philosophical Society." *Proceedings of the American Philosophical Society* 87, no. 3 (1943): 263–76.
———. "La Correspondance de Madame de Staël avec Jefferson." *Revue de Littérature Comparée* 2 (1922): 621–40.
———. *Trois amitiés françaises de Jefferson, d'après sa correspondance inédite avec Madame de Bréhan, Madame de Tessé et Madame de Corny.* Paris: Société d'Édition Les Belles Lettres, 1927.
Clagett, Martin. *Scientific Jefferson Revealed.* Charlottesville: University of Virginia Press, 2009.
Clark, Rex, and Oliver Lubrich, eds. *Transatlantic Echoes: Alexander von Humboldt in World Literature.* New York: Berghahn, 2012.
———, eds. *Cosmos and Colonialism: Alexander von Humboldt in Cultural Criticism.* New York: Berghahn, 2012.
Cogliano, Francis D. *Thomas Jefferson: Reputation and Legacy.* Charlottesville: University of Virginia Press, 2006.
Cohen, I. Bernard. *Science and the Founding Fathers: Science in the Political Thought of Thomas Jefferson, Benjamin Franklin, John Adams and James Madison.* New York and London: Norton, 1995.
Commager, Henry Steele. *The Empire of Reason: How Europe Imagined and America Realized the Enlightenment.* New York: Anchor, 1977.
Conkin, Paul K. "The Religious Pilgrimage of Thomas Jefferson." In *Jeffersonian Legacies,* edited by Peter S. Onuf, 19–49. Charlottesville: University Press of Virginia, 1993.
Coonen, Lester P., and Charlotte M. Porter. "Thomas Jefferson and American Biology." *BioScience* 26, no. 12 (1976): 745–50.
Cunningham, Noble E., Jr. *In Pursuit of Reason: The Life of Thomas Jefferson.* Baton Rouge: Louisiana State University Press, 1987.
Cushman, Gregory T. "Humboldtian Science, Creole Meteorology, and the Discovery of Human-Caused Climate Change in South America." In "Klima," special issue, *Osiris* 26 (2011): 16–44.
Cutright, Paul Russell. *Lewis and Clark: Pioneering Naturalists.* Lincoln and London: University of Nebraska Press, 1989.
Dassow Walls, Laura. "'Hero of Knowledge, Be Our Tribute Thine': Alexander von Humboldt in Victorian America." *Northeastern Naturalist* 8, no. 1 (2001): 129–33.

———. *The Passage to Cosmos: Alexander von Humboldt and the Shaping of America*. Chicago and London: University of Chicago Press, 2009.
———. "Rediscovering Humboldt's Environmental Revolution." *Environmental History* 10, no. 4 (2005): 758–60.
———. "The Search for Humboldt." *Geographical Review* 96 (2006): 473–77.
———. *Seeing New Worlds: Henry David Thoreau and Nineteenth-Century Natural Science*. Madison: University of Wisconsin Press, 1995.
Dettelbach, Michael. "Alexander von Humboldt between Enlightenment and Romanticism." *Northeastern Naturalist* 8, no. 1 (2001): 9–20.
Drouin, Jean-Marc. "Humboldt et la popularization des sciences." *La Revue, Musée des arts et métiers* 39/40 (2003): 45–63.
Dubois, Laurent. *Avengers of the New World: The Story of the Haitian Revolution*. Cambridge: Harvard University Press, 2004.
———. "The Haitian Revolution and the Sale of Louisiana; or, Thomas Jefferson's (Unpaid) Debt to Jean-Jacques Dessalines." In *Empires and Imagination: Transatlantic Histories of the Louisiana Purchase*, edited by Peter J. Kastor and Francois Weil, 93–116. Charlottesville and London: University of Virginia Press, 2009.
Dugatkin, Lee Alan. *Mr. Jefferson and the Giant Moose: Natural History in Early America*. Chicago and London: University of Chicago Press, 2009.
Dunn, Susan. *Sister Revolutions: French Lightning, American Light*. New York: Faber and Faber, 1999.
Egan, Clifford. "The United States, France, and West Florida, 1803–1807." *Florida Historical Quarterly* 47 (1968–69): 227–52.
Egerton, Douglas R. *Gabriel's Rebellion: The Virginia Slave Conspiracies of 1800 and 1802*. Chapel Hill: University of North Carolina Press, 1993.
Egerton, Frank N. "A History of the Ecological Sciences, Part 1: Early Greek Origins." *Bulletin of the Ecological Society of America* 82 (2001): 93–97.
———. "A History of the Ecological Sciences, Part 32: Humboldt, Nature's Geographer." *Bulletin of the Ecological Society of America* 90 (2009): 253–82.
Ewan, Joseph, and Nesta Dunn Ewan. *Benjamin Smith Barton: Naturalist and Physician in Jeffersonian America*. St. Louis: Missouri Botanical Garden Press, 2007.
Faak, Margot. *Alexander von Humboldts amerikanische Reisejournale: Eine Übersicht*. Berliner Manuskripte zur Alexander-von-Humboldt-Forschung, no. 25. Berlin: Alexander-von-Humboldt-Forschungsstelle, 2002.
Fiedler, Horst, and Ulrike Leitner. *Alexander von Humboldts Schriften: Bibliographie der selbständig erschienenen Werke*. Berlin: Akademie Verlag, 2000.
Finkelman, Paul. "Jefferson and Slavery: Treason against the Hopes of the World." In *Jeffersonian Legacies*, edited by Peter S. Onuf, 181–221. Charlottesville: University Press of Virginia, 1993.
Foner, Philip S. *Alexander von Humboldt on Slavery in the United States*. Berlin: Humboldt-Universität, 1984.
Foucault, Philippe. *Le Pêcheur d'orchidées: Aimé Bonpland, 1773–1858*. Paris: Seghers, 1990.

Fränzle, Otto. "Alexander von Humboldt's Holistic World View and Modern Inter- and Transdisciplinary Ecological Research." *Proceedings: Alexander von Humboldt's Natural History Legacy and Its Relevance for Today* 1 (2001): 57–90.

Friis, Hermann R. "Alexander von Humboldts Besuch in den Vereinigten Staaten von Amerika vom 20. Mai bis zum 30. Juni 1804." In *Alexander von Humboldt: Studien zu seiner universalen Geisteshaltung*, edited by Joachim H. Schultze, 142–95. Berlin: Walter de Gruyter, 1959.

———. "Baron Alexander von Humboldt's Visit to Washington." *Records of the Columbia Historical Society* 44 (1963): 1–35.

Gallatin, Albert. "Tabellarische Übersicht der Indianerstämme in den Vereinigten Staaten von Nordamerika, ostwärts von den Felsgebirgen (Stony Mountains), nach den Sprachen und Dialekten geordnet. 1826. Mitgetheilt von dem Freiherrn von Humboldt." *Hertha* 8 (1827): 328–34.

Geggus, David Patrick, ed. *Haitian Revolutionary Studies*. Bloomington and Indianapolis: Indiana University Press, 2002.

———. *The Impact of the Haitian Revolution in the Atlantic World*. Columbia: University of South Carolina Press, 2001.

———. "The Naming of Haiti." *New West Indian Guide* 71, no. 1/2 (1997): 43–68.

Gerbi, Antonello. *The Dispute of the New World: The History of a Polemic, 1750–1900*. Pittsburgh: University of Pittsburgh Press, 2010.

Gordon-Reed, Annette. *The Hemingses of Monticello: An American Family*. New York: Norton, 2008.

Gould, S. J. "Church, Humboldt and Darwin: The Tension and Harmony of Art and Science." In *Frederic Edwin Church*, edited by F. Kelly, Gould, J. A. Ryan, and D. Rindge, 94–107. Washington, D.C.: Smithsonian Institution Press, 1989.

Guha, Ramachandra. *Environmentalism: A Global History*. New York: Longman, 2000.

Guthorn, Peter J. "Kósciuszko as Military Cartographer and Engineer in America." *Imago Mundi* 29 (1977): 49–53.

Hailman, John. *Thomas Jefferson on Wine*. Jackson: University Press of Mississippi, 2009.

Hampe Martínez, Teodoro. "Carlos Montúfar y Larrea (1780–1816), el quiteño compañero de Humboldt." *Revista de Indias* 62, no. 226 (2002): 711–20.

Harris, Matthew L., and Jay H. Buckley, eds. *Zebulon Pike, Thomas Jefferson, and the Opening of the American West*. Norman: University of Oklahoma Press, 2012.

Hatch, Peter J. *The Fruits and Fruit Trees of Monticello*. Charlottesville and London: University Press of Virginia, 1998.

———. *The Gardens of Monticello*. Charlottesville: Thomas Jefferson Memorial Foundation, 1992.

———. *"A Rich Spot of Earth": Thomas Jefferson's Revolutionary Garden at Monticello*. New Haven: Yale University Press, 2012.

Helferich, Gerard. *Humboldt's Cosmos. Alexander von Humboldt and the Latin American Journey That Changed the Way We See the World.* New York: Gotham, 2004.

Herbst, Jürgen. "Thomas Jefferson und Wilhelm von Humboldt." In *Humboldt International: Der Export des deutschen Universitätsmodells im 19. und 20. Jahrhundert,* edited by Rainer Christoph Schwinges, 273–87. Basel: Schwabe, 2001.

Hernández González, Manuel. *Alejandro de Humboldt: Viaje a las Islas Canarias.* La Laguna: Francisco Lemus, 1995.

Hickey, Donald R. "America's Response to the Slave Revolt in Haiti, 1791–1806." *Journal of the Early Republic* 2, no. 4 (1982): 361–79.

———. "Timothy Pickering and the Haitian Slave Revolt: A Letter to Thomas Jefferson in 1806." *Essex Institute Historical Collections* 120 (1984): 149–63.

Hiepko, Paul. "Humboldt, His Botanical Mentor Willdenow, and the Fate of the Collections of Humboldt & Bonpland." *Bot. Jahrb. System* 126 (2006): 509–16.

Himmelfarb, Getrude. *The Roads to Modernity: The British, French, and American Enlightenments.* New York: Knopf, 2004.

Hoffmann, Léon-François, Frauke Gewecke, Ulrich Fleischmann, eds. *Haïti 1804—Lumières et ténèbres: Impact et résonances d'une revolution.* Madrid: Iberoamericana; Frankfurt am Main: Vervuert, 2008.

Humboldt, Alexander von. *Asie Centrale: Recherches sur les chaînes de montagnes et la climatologie compare.* 3 vols. Paris: Gide, 1843.

———. *Aus meinem Leben: Autobiographische Bekenntnisse.* Edited by Kurt-Reinhard Biermann. Munich: Beck, 1987.

———. *Briefe aus Amerika: 1799–1804.* Edited by Ulrike Moheit. Berlin: Akademie Verlag, 1993.

———. *Briefe von Alexander von Humboldt an Varnhagen von Ense aus den Jahren 1827 bis 1858.* Edited by Ludmilla Assing. Leipzig: F. M. Brodhaus, 1860

———. *Cosmos: A Sketch of the Physical Description of the Universe.* 4 vols. New York: Harper and Brothers, 1858.

———. *Cosmos: A Sketch of the Physical Description of the Universe.* Baltimore and London: John Hopkins University Press, 1997.

———. *Essay on the Geography of Plants.* Edited by Stephan T. Jackson. Chicago: University of Chicago Press, 2009.

———. *Florae fribergensis specimen.* Berolini: H. A. Rottmann, 1793.

———. "Fortschritte in der Kultur unter den Indiern Nordamerikas." *Hertha* 8 (1827): 320–328.

———. *The Island of Cuba.* New York: Derby and Jackson, 1856.

———. *Lateinamerika am Vorabend der Unabhängigkeitsrevolution: Eine Anthologie von Impressionen und Urteilen aus den Reisetagebüchern.* Vol. 5. Edited by Margot Faak. Berlin: Akademie Verlag, 1982.

———. *Mineralogische Beobachtungen über einige Basalte am Rhein.* Braunschweig: Schulbuchhandlung, 1790.

———. "Original Communication—Supplementary." *Literary Magazine and American Register* 2 (1804): 321–27.
———. *Political Essay on the Kingdom of New Spain.* 2 vols. New York: I. Riley, 1811.
———. *Political Essay on the Island of Cuba.* Edited by Vera M. Kutzinski and Ottmar Ette. Chicago and London: University of Chicago Press, 2011.
———. *Reise auf dem Rio Magdalena, durch die Anden und durch Mexiko.* Vol. 8. Edited by Margot Faak. 1986. Berlin: Akademie Verlag, 2003.
———. *Reise auf dem Rio Magdalena, durch die Anden und durch Mexiko.* Vol. 9. Edited by Margot Faak. Berlin: Akademie Verlag, 1990.
———. *Reise durch Venezuela.* Vol. 12. Edited by Margot Faak. Berlin: Akademie Verlag, 2000.
———. *Researches, Concerning the Institutions & Monuments of the Ancient Inhabitants of America, with Descriptions & Views of Some of the Most Striking Scenes in the Cordilleras.* 2 vols. London: Longman, 1814.
———. "Tablas geográfico-políticas del Reino de Nueva-España, en el año de 1803, que manifiestan su superficie, población, agricultura, fábricas, comercio, minas, rentas y fuerza militar. Por el Baron de Humboldt. Presentadas al Exmo. Señor Virey del mismo reino en enero de 1804." *Boletín de geografía y estadística* 1 (1869): 635–57.
———. "Über die Gestalt und das Klima des Hochlandes in der iberischen Halbinsel." *Hertha* 4 (1825): 5–23.
———. *Über die unterirdischen Gasarten und die Mittel ihren Nachtheil zu vermindern: Ein Beitrag zur Physik der praktischen Bergbaukunde.* Braunschweig: Friedrich Vieweg, 1799.
———. *Versuche über die chemische Zerlegung des Luftkreises und über einige andere Gegenstände der Naturlehre.* Braunschweig: Friedrich Vieweg, 1799.
———. *Versuche über die gereizte Muskel- und Nervenfaser nebst Vermuthungen über den chemischen Process des Lebens in der Thier- und Pflanzenwelt.* 2 vols. Posen: Decker und Compagnie; Berlin: Heinrich August Rottmann, 1797.
———. *Views of Nature or Contemplation on the Sublime Phenomena of Creation.* London: Henry G. Bohn, 1850.
———. *Views of Nature or, Contemplations on the Sublime Phenomena of Creation.* London: G. Bell & Daldy, 1872.
———. *Views of the Cordilleras and Monuments of the Indigenous Peoples of the Americas.* Edited by Vera M. Kutzinski and Ottmar Ette. Chicago: University of Chicago Press, 2012.
———. *Von Mexiko-Stadt nach Veracruz: Tagebuch.* Edited by Ulrike Leitner. Berlin: Akademie Verlag, 2005.
Humboldt, Alexander von, and Aimé Bonpland. *Personal Narrative of Travels to the Equinoctial Regions of the New Continent, during the Years 1799–1804.* 7 vols. London: Longman, 1814–29.
Humboldt, Alexander von, Aimé Bonpland, and Karl Sigismund Kunth. *Nova genera et species plantarum: Quas in peregrinatione ad plagam aequi-*

noctialem orbis novi collegerunt, descripserunt, partim adumbraverunt. 7 vols. Paris: Lutetiae, 1815–26.

Humboldt, Wilhelm von, and Jabbo Oltmanns. *Nivellement barométrique fait dans les régions équinoxiales du nouveau continent, en 1799–1804.* Paris: F. Schoell, 1809.

Hunt, Alfred N. *Haiti's Influence on Antebellum America: Slumbering Volcano in the Caribbean.* Baton Rouge: Louisiana State University Press, 1988.

Jackson, Donald Dean. *Thomas Jefferson and the Stony Mountains: Exploring the West from Monticello.* Urbana: University of Illinois Press, 1981.

Jahn, Ilse. "Alexander von Humboldt's Cosmical View on Nature and His Research Shortly before and Shortly after His Departure from Spain." In *Estudios de Historia das Ciencias e das Técnicas: VII Congreso de la Sociedad Española de Historia de las Ciencias y de las Técnicas,* edited by Mari Alvarez Lires, 31–40. 2 vols. Pontevedra: Diputación Provincial, 2001.

Jefferson, Thomas. "The Description of a Mould-Board of the Least Resistence, and of the Easiest and Most Certain Construction, Taken from a Letter to Sir John Sinclair, President of the Board of Agriculture at London." *Transactions of the American Philosophical Society* 4 (1799): 313–22.

———. *Jefferson's Extracts from the Gospels.* Edited by W. Dickinson Adams. Princeton: Princeton University Press, 1983.

———. *Jefferson's Memorandum Books: Accounts with Legal Records and Miscellany, 1767–1826.* Edited by James A. Bear and Lucia Stanton. 2 vols. Princeton: Princeton University Press, 1997.

———. *The Life and Selected Writings of Thomas Jefferson.* Edited by Adrienne Koch and William Peden. New York: Random House, 1993.

———. "A Memoir on the Discovery of Certain Bones of a Quadruped of the Clawed Kind in the Western Parts of Virginia." *Transactions of the American Philosophical Society* 4 (1799): 246–60.

———. *Notes on the State of Virginia.* Edited by William Peden. Chapel Hill: University of North Carolina Press, 1982.

———. *Notes on the State of Virginia.* Edited by Frank Shuffelton. New York: Penguin, 1999.

———. *The Papers of Thomas Jefferson.* Edited by Julian Boyd. 39 vols. to date. Princeton: Princeton University Press, 1950–2012–.

———. *The Papers of Thomas Jefferson Digital Edition.* Edited by Barbara B. Oberg and J. Jefferson Looney. Charlottesville: University of Virginia Press, Rotunda, 2008.

———. *Thomas Jefferson's European Travel Diaries.* Edited by James McGrath Morris and Persephone Weene. New York: Isidore Stephanus Son, 1987.

———. *Thomas Jefferson's Garden Book 1766–1824, with Relevant Extracts from Other Writings.* Annotated by Edwin Morris Betts. Charlottesville, Va.: Thomas Jefferson Memorial Foundation, 1999.

———. *The Writings of Thomas Jefferson.* Edited by Paul Leicester Ford. 10 vols. New York and London: G. P. Putnam's Sons, 1892–99.

———. *The Writings of Thomas Jefferson.* 20 vols. Edited by Andrew A.

Lipscomb and Albert Ellery Bergh. Washington, D.C.: Thomas Jefferson Memorial Association of the United States, 1903–7.
Johansen, Bruce Elliott. "Franklin, Jefferson and American Indians: A Study in the Cross-Cultural Communication of Ideas." Ph.D. diss., University of Washington, 1979.
Jordan, Winthrop D. *White over Black: American Attitudes toward the Negro, 1550–1812*. New York: Norton, 1968.
Kaminski, John P., ed. *Jefferson in Love: Love Letters between Thomas Jefferson and Maria Cosway*. Madison, Wisc.: Madison House, 1999.
Kaplan, Lawrence S. *Jefferson and France: An Essay on Politics and Political Ideas*. New Haven and London: Yale University Press, 1967.
Kastor, Peter J., ed. *The Louisiana Purchase: Emergence of an American Nation*. Washington, D.C.: CQ Press, 2002.
Kellner, Lotte. *Alexander von Humboldt*. London and New York: Oxford University Press, 1963.
Kendall, Joshua. *The Forgotten Founding Father: Noah Webster's Obsession and the Creation of an American Culture*. New York: Putnam, 2011.
Kennedy, Roger G. "Jefferson and the Indians." *Winterthur Portfolio* 27, no. 2/3 (1992): 105–21.
———. *Mr. Jefferson's Lost Cause: Land, Farmers, Slavery, and the Louisiana Purchase*. Oxford: Oxford University Press, 2003.
Kish, Georg, ed. *A Source Book in Geography*. Cambridge and London: Harvard University Press, 1978.
Köchy, Kristian. "Das Ganze der Natur—Alexander von Humboldt und das romantische Forschungsprogramm." *Humboldt im Netz* 3, no. 5 (2002). www.uni-potsdam.de/u/romanistik/humboldt/hin/hin5/koechy.htm.
Kohut, Karl. "Clavijero y las disputas sobre el Nuevo Mundo en Europa y América." *Destiempos* 3, no. 14 (2008): 52–81.
König, Clemens. "Willdenow, Karl Ludwig." In *Allgemeine Deutsche Biographie* (ADB), 43:252–54. Leipzig: Duncker and Humblot, 1898.
Krippendorff, Ekkehart. *Jefferson und Goethe*. Hamburg: Europäische Verlagsanstalt, 2001.
Kukla, Jon. *Mr. Jefferson's Women*. New York: Knopf, 2007.
———. *A Wilderness So Immense: The Louisiana Purchase and the Destiny of America*. New York: Knopf, 2003.
Kutzinski, Vera M., Ottmar Ette, and Laura Dassow Walls, eds. *Alexander von Humboldt and the Americas*. Berlin: Verlag Walter Frey 2012.
Labastida, Jaime, "Humboldt, México y Estados Unidos: Historia de una intriga." In *Atlas Geográfico y Físico del Reino de la Nueva España*, edited by Jaime Labastida and Charles Minguet, 131–47. Mexico: Siglo XXI, 2003.
Lack, Walther H. *Alexander von Humboldt and the Botanical Exploration of the Americas*. Munich, Berlin, London, and New York: Prestel, 2009.
Lane, Lawrence. "An Enlightened Controversy—Jefferson and Buffon." *Enlightenment Essays* 3, no. 1 (1972): 37–40.
Lange, Eugénie. "Aus dem Briefwechsel Alexander von Humboldts (1769–

1859) mit Thomas Jefferson (1743–1836)." *Societé Suisse des Americanistes* 18 (1959): 32–45.
Leitner, Ulrike, ed. *Alexander von Humboldt und Cotta: Briefwechsel.* Berlin: Akademie Verlag, 2009.
———. "'Anciennes folies neptuniennes!' Über das wiedergefundene Journal du Mexique à Veracruz aus den mexikanischen Reisetagebüchern A. v. Humboldts." *Humboldt im Netz* 3, no. 5 (2002). www.uni-potsdam.de/u/romanistik/humboldt/hin/ hin5/leitner.htm.
Lewis, James E. *The Louisiana Purchase: Jefferson's Noble Bargain?* Monticello Monograph Series. Chapel Hill, N.C.: Thomas Jefferson Foundation, 2003.
Lokke, Carl. "Jefferson and the Leclerc Expedition." *American Historical Review* 33 (1928): 322–28.
López Piñero, Jose Maria, and Thomas F. Glick. *El Megaterio de Bru y el Presidente Jefferson: Una relación insospechada en los albores de la paleontologia.* Valencia: Universidad Valencia/CSIC, 1993.
Macrory, Donald. *Nature's Interpreter: The Life and Times of Alexander von Humboldt.* Cambridge: Lutterworth Press, 2010.
Malone, Dumas. *Jefferson and His Time.* 6 vols. Boston: Little, Brown, 1948–82.
Manning, Susan, and Francis D. Cogliano, eds. *The Atlantic Enlightenment.* Aldershot, England, and Burlington, Vt.: Ashgate, 2008.
Martin, Edwin T. *Thomas Jefferson: Scientist.* New York: Collier, 1961.
Marvick, Elizabeth Wirth. "Thomas Jefferson and the Ladies of Paris." *Proceedings of the Annual Meeting of the Western Society for French History* 21 (1994): 81–94.
Mathewson, Kent. "Alexander von Humboldt's Image and Influence in North American Geography, 1804–2004." *Geographical Review* 96, no. 3 (2006): 416–38.
Matthewson, Tim. "George Washington's Policy towards the Haitian Revolution." *Diplomatic History* 3 (1979): 321–36.
———. "Jefferson and Haiti." *Journal of Southern History* 61, no. 2 (1995): 209–48.
———. "Jefferson and the Nonrecognition of Haiti." *Proceedings of the American Philosophical Society* 140, no. 1 (1996): 22–48.
———. "John Adams and the Independence of Haiti." Manuscript. Arlington, Va., 1994.
———. *A Proslavery Foreign Policy: Haitian-American Relations during the Early Republic.* Westport, Conn.: Praeger, 2003.
McDonald, Robert M. S., ed. *Light and Liberty: Thomas Jefferson and the Power of Knowledge.* Charlottesville: University of Virginia Press, 2012.
Mead, Robert Osborn. *Atlantic Legacy: Essays in American-European Cultural History.* New York: New York University Press, 1969.
Mergen, Bernard. *Snow in America.* Washington, D.C.: Smithsonian Institution Press, 1997.
Miller, Charles. *Jefferson and Nature: An Interpretation.* London: John Hopkins University Press, 1988.

Miller, John Chester. *The Wolf by the Ears: Thomas Jefferson and Slavery.* Charlottesville: University Press of Virginia, 1991.

Miller, Lillian B. "Charles Willson Peale as History Painter: The Exhumation of the Mastodon." *American Art Journal* 13, no. 1 (1981): 47–68.

Monreal, Sanz Marta, and Luis Álvarez Falcón. "Del racionalismo ilustrado a la sensibilidad romántica: La concepción singular del cambio de paradigma en la ciencia de Alexander von Humboldt." In *Estudios de Historia das Ciencias e das Técnicas: VII Congreso de la Sociedad Española de Historia de las Ciencias y de las Técnicas*, 2 vols., edited by Mari Alvarez Lires, 349–57. Pontevedra: Diputación Provincial, 2001.

Moore, Roy, and Alma Moore. *Thomas Jefferson's Journey to the South of France.* New York: Stewart, Tabori and Chang, 1999.

Morgan, Kathryn. *Jefferson and the Natural World: An Artist's Choice: The Catalogue of an Exhibition of the 250th Anniversary of the Birth of Thomas Jefferson.* Charlottesville: University Press of Virginia, 1993.

Nicholls, Michael L. *Whispers of Rebellion: Narrating Gabriel's Conspiracy.* Charlottesville: University of Virginia Press, 2012.

Nichols, Sandra. "Why Was Humboldt Forgotten in the United States?" *Geographical Review* 96 (2006): 399–415.

Nicolaisen, Peter. "Thomas Jefferson and Friedrich Wilhelm von Geismar: A Transatlantic Friendship." *Magazine of Albemarle County History* 64 (2006): 1–27.

Nicolson, Malcolm. "Humboldtian Plant Geography after Humboldt: The Link to Ecology." *British Journal for the History of Science* 29, no. 3 (1996): 290.

Nieto Olarte, Mauricio, Paola Castaño, and Diana Ojeda. "'El influjo del clima sobre los seres organizados' y la retórica ilustrada en el *Semanario del Nuevo Reino de Granada*." *Historia Crítica*, no. 30 (2005): 91–114.

Nieto Olarte, Mauricio. *Orden natural y orden social: Ciencia y política en el Semanario del Reyno de Granada.* Madrid: CSIC, 2007.

Nollendorfs, Cora Lee. "Alexander von Humboldt Centennial Celebrations in the United States: Controversies Concerning His Work." *Monatshefte* 80, no. 1 (1988): 59–66.

Onuf, Peter S. "To Declare Them a Free and Independent People: Race, Slavery, and National Identity in Jefferson's Thought." *Journal of the Early Republic* 18, no. 1 (1998): 1–46.

Oppitz, Ulrich-Dieter. "Der Name der Brüder Humboldt in aller Welt." In *Alexander von Humboldt, Werk und Weltgeltung*, edited by Heinrich Pfeiffer, 277–429. Munich: R. Piper, 1969.

Orrego González, Francisco. "Juan Ignacio Molina y la comprensión de la naturaleza del *finis terrae*: Un acercamiento desde la historia (cultural) de la ciencia." *Arbor* 187, no. 751 (2011): 961–76.

Palmer, Robert Roswell. "The Dubious Democrat: Thomas Jefferson in Bourbon France." *Political Science Quarterly* 72 (1957): 388–404.

Paret, Peter. "Jefferson and the Birth of European Liberalism." *Proceedings of the American Philosophical Society* 137, no. 4 (1993): 488–97.

Pattison, William D. "The Four Traditions of Geography." *Journal of Geography* 89, no. 5 (1990): 202–6.
Peale, Charles Wilson. *The Selected Papers of Charles Willson Peale and His Family*. Edited by Lillian B. Miller et al. New Haven: Yale University Press, 1983.
Peterson, Merrill D. "Thomas Jefferson and the French Revolution." *Tocqueville Review* 9 (1987/88): 15–25.
———. *Thomas Jefferson and the New Nation. A Biography*. New York: Oxford University Press, 1970.
Ponce, Esteban. "Fragmentos de un discurso no amoroso: Thomas Jefferson y la América Hispana. Una aproximación a las relaciones sur-norte." *Procesos: Revista Ecuatoriana de Historia* 30 (2009): 5–24.
Puig-Samper, Miguel Ángel. "Humboldt, un prusiano en la corte del Rey Carlos IV, 329–355. *Revista de Indias* 59, no. 216 (1999): 329–55.
Puig-Samper, Miguel Ángel, and Sandra Rebok. "Alexander von Humboldt y el relato de su viaje americano redactado en Filadelfia." *Revista de Indias* 62, no. 224 (2002): 69–84.
———. "Charles Darwin and Alexander von Humboldt: An Exchange of Looks between Two Famous Naturalists." *Humboldt im Netz* 9, no. 21 (2010). www.uni-potsdam.de/u/romanistik/humboldt/hin/hin21/puig-samper_rebok.htm.
———. *Sentir y medir: Alexander von Humboldt en España*. Aranjuez: Doce Calles, 2007.
———. "Un sabio en la meseta: El viaje de Alejandro de Humboldt a España en 1799." *Humboldt im Netz* 3, no. 5 (2002). www.uni-potsdam.de/u/romanistik/humboldt/hin/hin5/rebok.htm.
Ragosta, John A. *Religious Freedom: Jefferson's Legacy, America's Creed*. Charlottesville: University of Virginia Press, 2013.
Rebok, Sandra. "Alejandro de Humboldt y el modelo de la Historia Natural y Moral." *Humboldt im Netz* 2, no. 3 (2001). www.unipotsdam.de/u/romanistik/humboldt/hin/rebok-HIN3.htm.
———. *Alexander von Humboldt und Spanien im 19. Jahrhundert: Analyse eines wechselseitigen Wahrnehmungsprozesses*. Frankfurt: Vervuert, 2006.
———. "El arte al servicio de la ciencia: Alexander von Humboldt y la representación iconográfica de América." In publication on CD, *51° Congreso Internacional de Americanistas*, Santiago de Chile, July 2003.
———. "The Influence of Bernhard Varenius in the Geographic Works of Thomas Jefferson and Alexander von Humboldt." In *Bernhard Varenius (1622–1650)*, edited by Margarete Schuchard, 271–88. Leiden: Brill, 2008.
———. "A New Approach: Alexander von Humboldt's Perception of Colonial Spanish America as Reflected in His Travel Diaries." *Itinerario* 31, no. 1 (2007): 61–88.
———. "La Revolution de Haïti vue par deux personnages contemporains: Le scientifique prussien Alexander von Humboldt et l'homme d'état americain Thomas Jefferson." *French Colonial History* 10 (2009): 75–95.

———. "The Transatlantic Dialogue of the American Statesman Thomas Jefferson and the Prussian Traveller and Scientist Alexander von Humboldt." *Virginia Magazine of History and Biography* 116, no. 4 (2008): 329–69.

———. "Two Exponents of the Enlightenment: Transatlantic Communication by Thomas Jefferson and Alexander von Humboldt." In "Imagining the Atlantic World," special issue, *Southern Quarterly* 43, no. 4 (2006): 126–52.

———. *Una doble mirada: Alexander von Humboldt y España en el siglo XIX.* Madrid: CSIC, 2009.

Rice, Howard C., Jr. "Jefferson's Gift of Fossils to the Museum of Natural History in Paris." *Proceedings of the American Philosophical Society* 95, no. 6 (1951): 597–627.

Robinson, Joyce Henri. "An American Cabinet of Curiosities: Thomas Jefferson's 'Indian Hall at Monticello.'" *Winterthur Portfolio* 30, no. 1 (1995): 41–58.

Ronda, James P. *Jefferson's West: A Journey with Lewis and Clark.* Charlottesville, Va.: Thomas Jefferson Foundation, 2000.

Rupke, Nicolaas A. *Alexander von Humboldt: A Metabiography.* Frankfurt: Peter Lang, 2005.

Sachs, Aaron. *The Humboldt Current: Nineteenth-Century Exploration and the Roots of American Environmentalism.* New York: Viking, 2006.

———. "Humboldt's Legacy and the Restoration of Science." *World Watch* 8, no. 2 (1995): 29–39.

———. "The Ultimate Other: Post-Colonialism and Alexander von Humboldt's Ecological Relationship with Nature." *History and Theory* 42, no. 4 (2003): 111–35.

Sadosky, Leonard J., Peter Nicolaisen, Peter S. Onuf, and Andrew O'Shaughnessy, eds. *Old World, New World. America and Europe in the Age of Jefferson.* Charlottesville: University of Virginia Press, 2010.

Schneider, Richard J. *Thoreau's Sense of Place: Essays in American Environmental Writing.* Iowa City: University of Iowa Press, 2000.

Schneppen, Heinz. *Aimé Bonpland: Humboldts vergessener Weggefährte.* Berliner Manuskripte zur Alexander von Humboldt-Forschung 14. Berlin: Alexander-von-Humboldt-Forschungsstelle, 2000.

Schoenwaldt, Peter. "Alexander von Humboldt und die Vereinigten Staaten von Amerika." In *Alexander von Humboldt: Werk und Weltgeltung,* edited by Heinrich Pfeiffer, 431–82. Munich: R. Piper, 1969.

Schwarz, Ingo, ed. *Alexander von Humboldt und die Vereinigten Staaten von Amerika: Briefwechsel.* Berlin: Akademie Verlag, 2004.

———. "Alexander von Humboldts Bild von Latein- und Angloamerika im Vergleich." In *Nord u. Süd in Amerika: Gegensätze. Gemeinsamkeiten. Europäischer Hintergrund,* edited by Wolfgang Reinhard and Peter Waldmann, 2:1142–54. Freiburg: Rombach, 1992.

———. "Alexander von Humboldt's Visit to Washington and Philadelphia, His Friendship with Jefferson, and His Fascination with the United

States." In "Proceedings: Alexander von Humboldt's Natural History Legacy and Its Relevance for Today," special issue, *Northeastern Naturalist* 1 (2001): 43–56.

———. "Alexander von Humboldt—Socio-political Views of the Americas." In *Ansichten Amerikas: Neuere Studien zu Alexander von Humboldt,* edited by Ottmar Ette and Walther L. Bernecker. Frankfurt am Main: Vervuert, 2001.

———. "From Alexander von Humboldt's Correspondence with Thomas Jefferson and Albert Gallatin." Berliner Manuskripte zur Alexander-von-Humboldt-Forschung 2 (1991): 1–20.

———. "'Shelter for a Reasonable Freedom' or Cartesian Vortex." In *Debates y perspectivas: Alejandro de Humboldt y el mundo hispánico,* no. 1, edited by Miguel Ángel Puig-Samper, 169–82. Madrid: Fundación Histórica Tavera, 2000.

Sellers, Charles Coleman. *Mr. Peale's Museum: Charles Willson Peale and the First Popular Museum of Natural Science and Art.* New York: Norton, 1980.

Shapley, Harlow. "Notes on Thomas Jefferson as a Natural Philosopher." *Proceedings of the American Philosophical Society* 87, no. 3 (1943): 234–37.

Sheehan, Bernard W. *Seeds of Extinction: Jeffersonian Philanthropy and the American Indian.* New York: Norton, 1974.

Sheridan, Eugene R. *Jefferson and Religion.* Monticello Monograph Series. Charlottesville, Va,: Thomas Jefferson Memorial Foundation, 1998.

Shugart, H. H., and F. I. Woodward. *Global Change and the Terrestrial Biosphere: Achievements and Challenges.* Hoboken, N.J.: Wiley-Blackwell, 2011.

Sidbury, James. "Saint Domingue in Virginia: Ideology, Local Meanings, and Resistance to Slavery, 1790–1800." *Journal of Southern History* 63, no. 3 (1997): 531–52.

Sloan, Phillip R. "The Buffon-Linnaeus Controversy." *Isis* 67, no. 3 (1976): 356–75.

Smith, Margaret Bayard. *The First Forty Years of Washington Society.* Edited by Gaillard Hunt. New York: Scribner's Sons, 1906.

Smith Barton, Benjamin. "A Botanical Description of the Podophyllum Diphyllum." *Transactions of the American Philosophical Society* 3 (1793): 334–47.

Sofka, James R. *Metternich, Jefferson, and the Enlightenment: Statecraft and Political Theory in the Early Nineteenth Century.* Madrid: CSIC, 2011.

Sowerby, E. Millicent, ed. *Catalogue of the Library of Thomas Jefferson.* 4 vols. Charlottesville: University Press of Virginia, 1983.

Stagg, John C. A. *Borderlines in Borderlands: James Madison and the Spanish-American Frontier, 1776–1821.* New Haven: Yale University Press, 2009.

Stanton, Lucia. "Better Tools for a New and Better World. Jefferson Perfects the Plow." In *Old World, New World,* edited by Leonard J. Sadosky, Peter Nicolaisen, Peter S. Onuf and Andrew O'Shaughnessy, 200–222. Charlottesville: University of Virginia Press, 2010.

———. *Free Some Day: The African-American Families of Monticello.* Charlottesville, Va.: Thomas Jefferson Foundation, 2000.

———. *Slavery at Monticello*. Charlottesville, Va.: Thomas Jefferson Foundation, 1996.
———. *"Those Who Labor for My Happiness": Slavery at Thomas Jefferson's Monticello*. Charlottesville: University of Virginia Press, 2012.
Stearn, William Thomas, ed. *Humboldt, Bonpland, Kunth and Tropical American Botany*. Lehre: Cramer, 1968.
Stoddard, Richard Henry. *The Life, Travels and Books of Alexander von Humboldt*. With an introduction by Bayard Taylor. New York: Rudd and Carleton, 1859.
Terra, Helmut de. "Alexander von Humboldt's Correspondence with Jefferson, Madison and Gallatin." *Proceedings of the American Philosophical Society* 103 (1959): 783–806.
———. *Humboldt: The Life and Times of Alexander von Humboldt. 1769–1859*. New York: Knopf, 1955.
———. "Motives and Consequences of Alexander von Humboldt's Visit to the United States (1804)." *Proceedings of the American Philosophical Society* 104, no. 3 (1960): 314–16.
———. "Studies of Documentation of Alexander von Humboldt." *Proceedings of the American Philosophical Society* 102, no. 2 (1958): 136–41, and 102, no. 6 (1958): 560–56.
Théodoridès, Jean. "Les séjour aux Etats-Unis de deux savants européens de XIXe siècle: Alexander von Humboldt et Victor Jacquemont." *Archives Internationales d'histoire des Sciences* 16, no. 64 (1963): 287–304.
Thompson, Joseph P., Francis Lieber, Charles P. Daly, A. D. Bache, and George Bancroft Guyot. "Proceedings: Alexander von Humboldt Commemoration." *Journal of the American Geographical and Statistical Society* 1, no. 8 (1859): 225–46.
Thomson, Keith. *Jefferson's Shadow: The Story of His Science*. New Haven: Yale University Press, 2012.
———. *The Legacy of the Mastodon: The Golden Age of Fossils in America*. New Haven and London: Yale University Press, 2008.
———. *A Passion for Nature. Thomas Jefferson and Natural History*. Monticello Monograph Series. Monticello: Thomas Jefferson Foundation, 2008.
Valsania, Maurizio. *The Limits of Optimism: Thomas Jefferson's Dualistic Enlightenment*. Charlottesville: University of Virginia Press, 2011.
Wallace, Anthony F. C. *Jefferson and the Indians: The Tragic Fate of the First Americans*. Cambridge: Belknap Press of Harvard University Press, 1999.
Wassermann, Felix M. "Six Unpublished Letters of Alexander von Humboldt to Thomas Jefferson." *Germanic Review* 29 (1954): 191–200.
Watts, George B. "Thomas Jefferson, the 'Encyclopédie' and the 'Encyclopédie méthodique.'" *French Review* 38, no. 3 (1965): 318–25.
West, Susan. "Jefferson as Scientist." *Science News* 119, no. 19 (1981): 298–99.
Wills, Garry. *Negro President: Jefferson and the Slave Power*. Boston: Houghton Mifflin, 2005.
Wilson, Gaye. "'Behold Me at Length on the Vaunted Scene of Europe': Thomas Jefferson and the Creation of an American Image Abroad." In *Old*

World, New World, edited by Leonard J. Sadosky, Peter Nicolaisen, Peter S. Onuf, and Andrew O'Shaughnessy, 155–78. Charlottesville: University of Virginia Press, 2010.

———. "Jefferson, Buffon, and the Mighty American Moose." *Monticello Newsletter* 13, no. 1 (2002).

Wistar, Caspar. "A Description of the Bones Deposited by the President in the Museum of the Society." *Transactions of the American Philosophical Society* 4 (1799): 526–31.

Worster, Donald. *Nature's Economy: A History of Ecological Ideas*. Cambridge: Cambridge University Press, 1994.

Wulf, Andrea. *Brother Gardeners: Botany, Empire, and the Birth of an Obsession*. New York: Knopf, 2009.

Yacou, Alain, ed. *Saint-Domingue espagnol et la révolution nègre d'Haïti (1790–1822): Commémoration du bicentenaire de la naissance de l'État d'Haïti (1804–2004)*. Paris: Karthala, 2007.

Zeuske, Michael, "Alexander von Humboldt y la comparación de las esclavitudes en las Américas." *Humboldt im Netz* 7, no. 11 (2005). www.uni-potsdam.de/u/romanistik/humboldt/in/hin11/inh_zeuske_1.htm.

———. "Humboldt, esclavitud, autonomismo y emancipación en las Américas, 1791–1825." In *Alexander von Humboldt: La estancia en España y su viaje americano*, edited by Mariano Cuesta Domingo and Sandra Rebok, 257–77. Madrid: Real Sociedad Geográfica, CSIC, 2007.

Zuckert, Michael P. "Self-Evident Truth and the Declaration of Independence." *Review of Politics* 49, no. 3 (1987): 319–39.

Index